Toyoaki Nishida, Lakhmi C. Jain, and Colette Faucher (Eds.)

Modeling Machine Emotions for Realizing Intelligence

Smart Innovation, Systems and Technologies 1

Editors-in-Chief

Prof. Robert James Howlett
KES International
85 Victoria Road
Shoreham-by-sea
West Sussex
BN43 5WR, UK
E-mail: rjhowlett@kesinternational.org

Prof. Lakhmi C. Jain
School of Electrical and Information Engineering
University of South Australia
Adelaide, Mawson Lakes Campus
South Australia SA 5095
Australia
E-mail: Lakhmi.jain@unisa.edu.au

Further volumes of this series can be found on our homepage: springer.com

Vol. 1. Toyoaki Nishida, Lakhmi C. Jain, and Colette Faucher (Eds.)
Modeling Machine Emotions for Realizing Intelligence, 2010
ISBN 978-3-642-12603-1

Toyoaki Nishida, Lakhmi C. Jain, and Colette Faucher
(Eds.)

Modeling Machine Emotions for Realizing Intelligence

Foundations and Applications

Springer

Prof. Toyoaki Nishida
Dept. of Intelligence Science and Technology
Graduate School of Informatics,
Kyoto University
Yoshida-Honmachi
Sakyo-ku
Kyoto 606-8501
Japan
E-mail: nishida@i.kyoto-u.ac.jp

Prof. Lakhmi C. Jain
SCT-Building
University of South Australia
Adelaide
Mawson Lakes Campus
South Australia
Australia
E-mail: Lakhmi.jain@unisa.edu.au

Dr. Colette Faucher
Université Paul Cézanne
Polytech'Marseille
Avenue Escadrille Normandie-Niemen,
13397 Marseille Cedex 20
France

ISBN 978-3-642-12603-1 e-ISBN 978-3-642-12604-8

DOI 10.1007/978-3-642-12604-8

Smart Innovation, Systems and Technologies ISSN 2190-3018

Library of Congress Control Number: 2010924913

Typesetting: Scientific Publishing Services Pvt. Ltd., Chennai, India.

Printed on acid-free paper

9 8 7 6 5 4 3 2 1

springer.com

Preface

Emotion connects the thought to the body, which is a magnificent biological device for sensing and affecting the world. The thought controls the body through emotions. The body affects the thought through emotions. Through this mechanism, the thought allows the agent to behave intelligently in the complex world filled with a huge amount of dynamic information. The emotion maps a flux of information into a space which the agent is familiar with, enabling her/him to associate ongoing events with past experiences which help to reduce complexity by providing with a nominal solution.

Recent findings in brain science suggest that mirror neurons map visual signals into motor signals for the body. This mechanism might permit one to experience the emotion of the other agent just by feeling the motor signals caused by mirror neurons as a result of visual stimuli caused by the other agent's emotional behaviors. In particular, it might play a significant role in invoking empathy in a social situation.

It may not be hard to think about what might happen to emotion-less machines. The emotion-less machines may not be able to accumulate experiences to avoid serious failures. They may not be able to communicate with the humans in an empathetic way.

This research book aims to present recent progress in modeling and synthesizing emotional intelligence. The book consists of ten chapters. The first chapter overviews discussions in this volume. It describes major concepts and issues underlying primitive machineries, appraisal theories, primary and secondary emotions, and image-based thought processes. It discusses how emotional engines might be incorporated into an intelligent system.

The nine chapters following this introductory chapter span a wide spectrum of issues concerning emotional models, ranging from a psychological model of priming timing effects to applications to emotionally intelligent agents. Those chapters may be classified into three groups. The first group (chapters 2-3) is concerned with physiological and psychological models of emotions. They aim at building detailed quantitative models based on physiological measurement. The second group (chapters 4-8) addresses a computational model of emotion. They are about building emotionally intelligent agents, either intelligent virtual humans or robot, that can act wisely or engage in social interaction with people. The third group (chapters 9-10) addresses long-term, high-level emotions involving such issues as episodic memory models or compassionate intelligence.

In Chapter 2, Huber and Winkielman proposes a three-layered neural network model to account for a mechanism for valence judgments in affective processing consisting of priming, identification, attribution, and habituation. Huber and Winkielman suggest that the priming and habituation are automatically performed below the conscious level, while identification and attribution are deliberately with the strategic mechanism. The model is general enough to qualitatively reproduce dynamical aspects in performing perceptual identification tasks, assimilation and contrast effects in particular.

In Chapter 3, Ohmoto, Miyake and Nishida addresses a method of estimating the affective status of the communication partners solely from visual information. They introduce the notion of I-measures to characterize the affectively positive status of the communication partners, such as involvement, enjoyment, or excitement. They distinguish the intrinsic I-measure resulting from one's own emotional reaction to the stimulus from the extrinsic I-measure that is induced by the affective reactions of other participants. They experimentally show that both categories of I-measure can be detected from physiological measures (SCR and respiration). They also show that there is a positive correlation between the physiological and visual measure (body movement), so that about 70% of I-measure can be visually detected by carefully controlling the threshold.

In Chapter 4, emphasis is placed on the learning low level emotions. Mohammad and Nishida point out that utilization of suitable interaction protocol for body alignment and synchrony is critical to maintain positive emotional response of the interaction partner. They propose a mechanism modeling approach to induce the hypothesized underlying mechanisms and structures that can generate the observed behaviors. This approach is contrasted to behavior modeling approaches to reproduce the observed behaviors regardless of the internal processes or structures of the agent under consideration. They show a computational framework L_iEICA and three-staged algorithm for learning interaction protocol, consisting of autonomous learning of basic interactive acts (interaction bubbling), autonomous learning of interaction structure (protocol learning), and adaptation through interaction. Based on the gaze controllers building experiments, they have shown that the mechanism modeling approach provides more acceptable behaviors according to the participants' subjective evaluations.

In Chapter 5, Kubota and Wakisaka use an emotional model to build partner robots. They model emotions at three levels. On the first level, they characterize emotion as an intense short-term internal state of an agent based on perceptual information. On the second level, they model feeling as an accumulated value of emotions. On the third level, they define mood as a long-term state by accumulating feelings. The mood is modeled to govern changes in feelings. The introduction of the emotional model is reported to contribute to realize experience-based, location-dependent episodic memory as well as to increase communicability with the partner robots.

In Chapter 6, Zhang et al present a method for producing emotional facial expression from a parametric representation of an emotion state. Emotional states are represented by combining three parameters in the PAD (Pleasure-Arousal-Dominance) model that serves as a high-level emotion descriptor. Emotional facial expressions are described with the MPEG-4 FAP (Facial Animation Parameters) that specifies face model deformation. A unique feature about this framework is that it introduces the PEP (Partial Expression Parameters) as an intermediate representation for a facial motion pattern associated with a specific emotional state. The PAD-PEP mapping function and the PEP-FAP translation function are trained on a relatively small facial expression database.

In Chapter 7, Ohashi, Huang and Nishida consider that employing appropriate emotion model is critical to make embodied conversational agents attentive. They use a state transition model to simulate the dynamic mental process during the interaction with the user. The initial structure of the state transition model was designed manually based on a preliminary experiment. The behavior of a human operator in the preliminary WOZ experiments is recorded and used to train SVM (Support Vector Machine) to produce transition probabilities. The emotion engine is incorporated into a quiz agent. As a result of the GNAT (Go/No-go Association Task), it was shown that the quiz agent was judged to exhibit natural behaviors.

In Chapter 8, Kiselev et al argue for incorporating emotional intelligence into a virtual technical support engineer (VTSE), which is an embodied conversational agent that can help the user engage in natural interactions to receive technical support. The role of emotional intelligence in this task is to comprise a ground that permits VTSE to follow efficient and robust communication with the user. The architecture of VTSE is given which integrates emotional intelligence with function / behavior markup language for controlling interactions. A methodology for evaluating the effect of emotional intelligence is given by combining a conventional questionnaire-based method and implicit association test.

In Chapter 9, Neviarouskaya, Prendinger, and Ishizuka address affective text classification and its applications. They present Affect Analysis Model consisting of symbolic cue analysis, syntactic structure analysis, word-level analysis, phrase-level analysis, and sentence-level analysis. It features analysis of nine emotions at the individual sentence level, the ability to handle the evolving nature of the language for online communication, rich database of linguistic expressions concerning emotions, effective representation of affective features, powerful method for analyzing linguistic structure, and emotion intensity estimation. The experimental evaluation revealed that the Affect Analysis Model can classify emotions with 70-80% accurately. A couple of applications are shown to emotionally intelligent computer-mediated communication: one for instant messaging and another for Second Life.

In Chapter 10, Kasap and Magnenat-Thalmann discuss the role of episodic memory to enable long-term emotional interactions. In order to realize long-term believability, the episodic memory should be able to remember past knowledge and retrieve it in future through adequate methods of encoding, storage and retrieval. In order to maintain social relationship between the user and the agent, emotional expressions should be created and interpreted coherently.

We are grateful to the authors and the reviewers for their vision and efforts. We would like to thank Springer-Verlag and the publishing team for their assistance during the production phase of this book.

Toyoaki Nishida
Lakhmi C. Jain
Colette Faucher

Table of Contents

Chapter 1
Modeling Machine Emotions for Realizing Intelligence – An Introduction –

Toyoaki Nishida

Graduate School of Informatics, Kyoto University
Sakyo-ku, Kyoto 606-8501, Japan
nishida@i.kyoto-u.ac.jp
http://www.ii.ist.i.kyoto-u.ac.jp/~nishida/

Abstract. This chapter provides an overview of the materials contained later in this volume. I highlight major aspects of emotions and consider how they may be realized in a computational architecture. I consider issues related to primitive machineries, appraisal theories, primary and secondary emotions, and image-based thought processes. I discuss how emotional engines are incorporated into an intelligent system.

1 Introduction

Emotions are thought to be essential to intelligence. Is logical or rational thinking sufficient to produce intelligence or communicate intentions? Early AI systems did not employ emotions, and worked alone using logical reasoning. What was lacking in this approach?

An emotion-less machine intelligence might be brittle. Emotion-less machines may suffer in that they may employ infinite loops for searching. They are unable to recognize a tacit danger even though we can easily distinguish it.

An emotion-less machine intelligence may cause communication difficulties unless it can recognize and understand how the user may be tired and bored reacting with an agent. The humans will treat them only as a tool and not use them as a partner, which might not what the creator of the program had intended. The lack of the intentional stance will prevent a social relationship between the user and the machine from developing. It may be pretty hard to create a machine that appears to possess emotions even if the code is explicitly written to simulate emotions. For example, a simple mobile robot that is programmed to detect and approach red objects may not appear to function as if it *liked* red objects, except for a while.

Although it might be a bit too hard to say that possessing an emotional system is a sufficient condition for a machine to be intelligent, some kind of emotional system is deemed necessary for a system to be intelligent.

Similarly, an emotional system appears to be mandatory for a system to be able to communicate with the user in a natural fashion. Should future machines be designed to allow fluent communication with people, the ability to recognize and express emotions are inherent in human-like high level communications. It is hard

T. Nishida et al. (Eds.): Modeling Machine Emotions for Realizing Intelligence, SIST 1, pp. 1–15.
springerlink.com

to think about fluent inter-human communication without these abilities. It appears that the machines need to possess and behave with an appropriate system of emotions in order to be intelligent.

A number of valuable suggestions have been made about emotional intelligence in the past two decades. Daniel Goleman discussed how emotion is central to intelligence [1], [2]. Bechara showed that decision-making depends on emotion that patients impaired in amygdala and ventromedial prefrontal cortex have problems in decision-making when gambling [3]. In Descartes' Error, Antonio Damasio pointed out how intelligence may incur danger without emotions [4]. Ortony, Clore, and Collins showed how emotions are associated with valenced responses to goal-oriented events [5]. Rosalin Pickard proposed affective computing as a principle to design emotional intelligence, such as an affective tutor that can change the teaching mode according to the affective status of the learner [6].

After the initial phase of provocative and ideological discussions about emotional intelligence, we enter the phase of elaborating on the idea of emotional intelligence from technical point of view. The focus of discussions shifts from the principles of emotional intelligence to practical issues such as building computational model of emotions, enabling technologies and applications of emotional intelligence.

The role of emotional intelligence may become clearer by drawing the diagram of the whole intelligent system and then discussing how emotion intelligence helps the intelligent system behave in the world. We gradually shift the focus to the architecture of sophisticated emotional intelligence, starting from the most primitive one.

The intention of this chapter is to draw a big picture suggesting how emotional intelligence might be embedded into an intelligent system, how to identify key issues, and to indicate how they are connected with each other.

2 The Emotion-Less Primitive Model

It might be helpful to begin discussions about machines without emotions. Fig. 1 shows two typical architectures for machines which do not have explicit components for emotion. The one on the left shows the most primitive. It suggests that the motor is controlled by the sensor input, and the real world is updated by the motor. The one on the right shows the production system architecture, which is most typically used for knowledge-based systems particularly in the early days of artificial intelligence research. From the viewpoint of the discussions in this volume, a prominent feature of this architecture is that it introduces a layer of information representation on which the "brain" of the intelligent machine can take hypothetical actions. These may include planning the future actions without affecting to the real world.

The architectures shown in Fig. 1 are universal in the sense that they allow one to simulate almost any processes so long as they are computational. However, they are weak as a conceptual framework of intelligent machine, as they have no

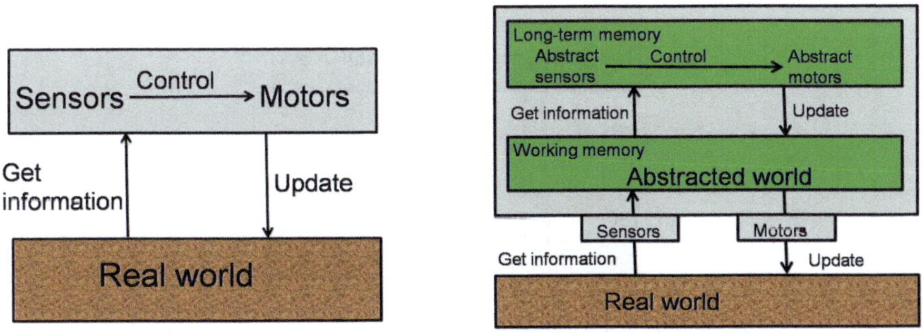

Fig. 1. Typical computational architectures for emotion-less intelligence

provision to prevent inappropriate behaviors. These would include such things as endless pursuit for the goal without doubt or failure in response to urgent problems. An appropriate conceptual architecture for intelligent machines should display commitment to intelligent behavior.

3 The Primitive Emotion Model

Braitenberg pointed out that even a simple machinery can produce a feeling that it exhibits an emotion [7]. Consider two simple two sensor-controlled vehicles as shown in Fig. 2. In each of the vehicles, as the sensor receives the stronger signal from the light source, it will produce the stronger motor torque. The difference between the two vehicles is the way in which the sensors are connected to the motors. The left will move away from the light source, as its right sensor receives a stronger light than the left as it is closer to the light source than the other. It then produces a stronger torque in the right wheel than the left. Consequently, the vehicle on the left may appear to *fear* the light. By contrast, the vehicle on the right will move toward the light source, as if it *liked* the light.

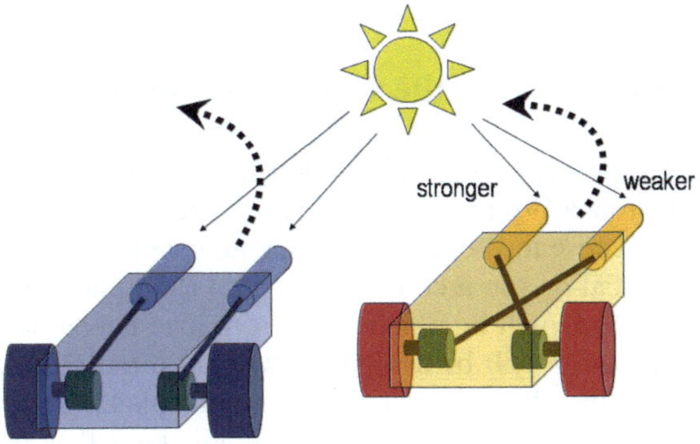

Fig. 2. Two kinds of "emotional" vehicles [7]. The vehicle on the left may appear to fear the light, while that on the right may appear as if it liked the light.

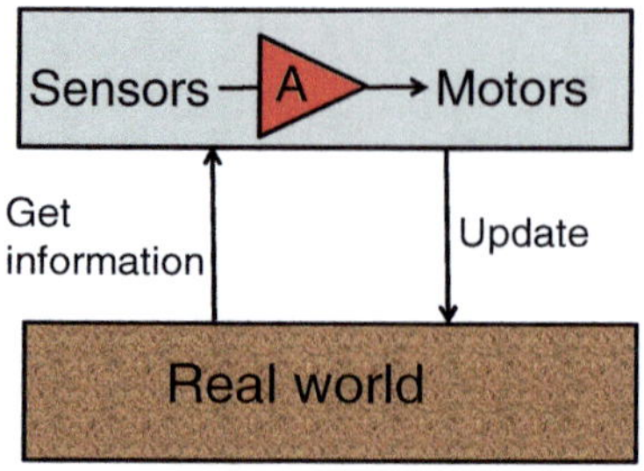

Fig. 3. A computational model for the Breitenberg's vehicles

From computational points of view, the mechanism of Breitenberg's vehicles can be abstracted and shown as the model in Fig. 3. This results by adding to the first model (Fig. 1) one component (module A) that maps sensor information to the motors. The mapper then produces an appraisal from a collection of sensory inputs. The mapper could be hardwired as it will not change during the lifetime of the creature. As discussed, the difference in the mapper could result in different personal traits. The mapper could be the result from the evolutionary process of the class of the agents. This could occur over many generations, accumulating experiences of that embodiment.

The Breitenberg's vehicles suggest a number of interesting conjectures. For example, the emotion might result from a relatively simple biological mechanism even though it may be more complex than Breitenberg's vehicles. Such a conjecture is over-generalization and emotions may be attributed only to the observer's cognition. Intellectual observers might only attribute emotions on to this simple machinery. The model shown in Fig. 3 fails to explain other issues which are related to the emotions. In particular, it does not indicate how such a low-level addition could be incorporated into higher levels of intelligence. More sophisticated treatments will be needed to convert appraisals from primitive "emotion" engine to account for and predict behaviors.

4 The Emotion Model at the Cognitive Level

Various attempts were made, which included those in the ALIVE project [8], or believable agents project [9], to build artificial creatures that are convincing in the sense that they may cause the observer to suspend disbelief.

In these agents, goal-driven architecture is used to implement these elementary artificial creatures. Plans will be composed to produce behavior sequences to satisfy these goals. Intention will be associated with an internal state like "I am currently at this stage of the plan and will commit to this plan."

Emotions may be associated with various mental states that arise when seeking a goal. For example, the actor may be deeply *disappointed* if s/he finds an event

Consequences of events	[+]					"pleased"
	[−]					"displeased"
	Focus	Consequences for others	[+]	Desirable for other		"happy-for"
				Undesirable for other		"gloating"
			[−]	Desirable for other		"resentment"
				Undesirable for other		"pity"
		Consequences for self	[+]	Prospects relevant		"hope"
					Confirmed	"satisfaction"
					Disconfirmed	"disappoint-ment"
				Prospects irrelevant		"joy"
					Attributed to self (agent)	"gratification"
					Attributed to other agent	"gratitude"
			[−]	Prospects relevant		"fear"
					Confirmed	"fears-confirmed"
					Disconfirmed	"relief"
				Prospects irrelevant		"distress"
					Attributed to Self (agent)	"remorse"
					Attributed to other agents	"anger"
Actions of agents	[+]					"approving"
	[−]					"disapproving"
	Focus	Self agent	[+]			"pride"
			[−]			"shame"
		Other agent	[+]			"admiration"
			[−]			"reproach"
Aspects of Objects	[+]					"liking"
	[−]					"disliking"
	Attraction	[+]				"love"
		[−]				"hate"

Fig. 4. The OCC (Ortony, Clore, and Collins) model [5]

suggesting, for example, that a rescue ship has left without finding her/him when s/he has been looking for help in the sea. This state, however, will change to that of *being-relieved* when it finds a helicopter approaching with a rescue net.

Cognitive appraisal theories give theoretical accounts on how emotions may be associated with the events that the agent encounters at each of the stages. The events are appraised and emotions with positive or negative senses result.

Ortony, Clore, and Collins showed a comprehensive model, which is often referred to as the OCC model, for accounting how emotions are associated with valenced responses to goal-oriented events [5]. The OCC model is shown in Fig. 4. For example, if the consequences of an event is positive, the resulting emotion might be designated as "pleased" in general. Otherwise, it is "displeased". A more detailed appraisal may arise by taking into account other aspects of such as occurrences, whether the focus is the consequences for self, prospects are irrelevant, and the consequences are attributed to other agents. The resulting emotion might be "gratitude" or "anger", depending on whether the consequences are positive or not. The OCC model is prominent, for it suggests the logic of the interface between the emotional engine and the whole intelligence system.

Some researchers have attempted to represent the varieties of emotions by combinations of "basic" emotions. Russell and Mehrabian presented a three-factor theory of emotion, where they showed that emotional states are represented by the three bipolar dimensions of pleasure-displeasure, degree of arousal, and dominance-submissiveness (PAD) [10]. Mehrabian presents the PAD temperament model in which a comprehensive list of mapping different emotions into the PAD space is demonstrated [11]. Fig. 5 shows how different kinds of emotion can be mapped into the PAD space. For example, "exuberant" (P=+, A=+, D=+) is placed as the opposite extreme of "bored" (P=−, A=−, D=−), "docile" (P=+, A=−, D=−) as the opposite extreme of "hostile" (P=−, A=+, D=+), for example.

The OCC model or other appraisal theories might be incorporated into the cognitive architecture as shown in Fig. 6. It is different from the model shown in Fig. 3. Here, the sensory inputs are temporally stored in the working memory in a

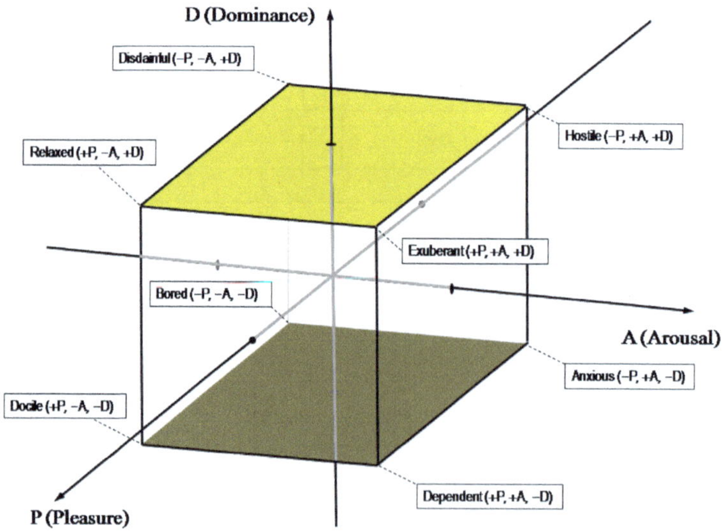

Fig. 5. PAD (Pleasure-Arousal-Dominance) model [10]

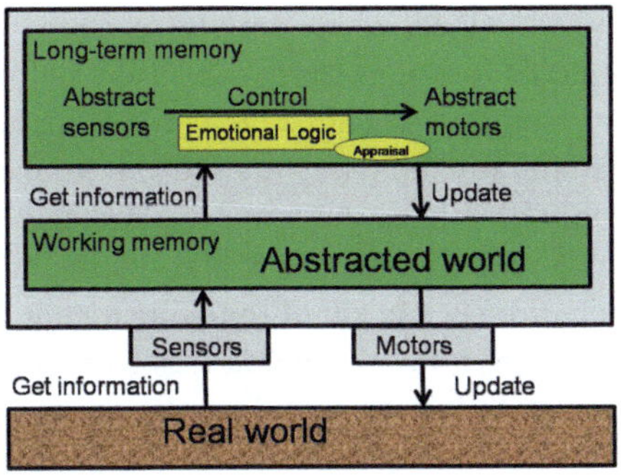

Fig. 6. A computational model that embodies the OCC (Ortony, Clore, and Collins) model

form more abstract than physical signals so that the appraisals can be handled at a level based on what may be called Emotional Logic. The question remains why emotions need to arise as a result of events. This is necessary to change the mood of the agent, from strained to relaxed. The OCC model does not suggest the mechanism of associating emotions with events. It may not be enough to directly couple emotions with events. Such a direct coupling does not specify how the transitions from one state to another can occur in a reasonable way. It does not distinguish primary and secondary emotions either.

5 Primary and Secondary Emotions

Antonio Damasio discussed how emotion is built on the brain and the body [4]. For example, Damasio suggests that a forced laugh comes from a mechanism different to a natural smile. Muscles for natural smile are controlled by the limbic cortex and probably basal nuclei. Those for a forced smile is controlled by the motor cortex and pyramidal system.

Among others, Damasio pointed out that the emotion system consists of at least two different systems: primary and secondary emotions. Primary emotion is manifested as a quick reflective reaction to a stimulus, such as pain arising from being hit by a hard object. The secondary emotion results from subsequent deliberation. Primary emotion is useful for an actor to notice an immediate danger as it happens even though the actor may be occupied by other businesses. A quick reaction may occur in order to temporally avoid the danger and to seek an appropriate reaction. In contrast, the secondary emotion is more logical and deliberate. Past experiences might be investigated in order to find a solution to resolve the cause of the primary emotion.

Damasio suggests a brain mechanism for the primary and secondary emotions. The primary emotion uses amygdala. When the stimulus arrives from sensors, they are directly sent to amygdala, where various reactions will be created such as internal reactions, muscular reactions, reactions in the automatic nervous system, and reactions in nucleus and hypothalamic area producing neurotransmitter. These reactions take place without invoking the brain functions and the response is rapid. In contrast, for secondary emotions, the brain receives and interprets signal. As a result, frontal cortex will be activated, and will also stimulate the amygdala. Thus, a new system utilizes an old mechanism of initiating emotional behavior.

The block diagram shown in Fig. 7 summarizes Damasio's discussion. Sensory inputs are sent in parallel to amygdala and the brain cognitive processes. Upon the receipt of the signals, amygdala creates varieties of physiological responses in the body. This corresponds to the primary emotions. The signals are also interpreted by the cognitive functions of the brain. This may trigger the ventromedial prefrontal cortex (VMF) to send signals to the amygdala and create effects similar to those caused by sensory inputs. The entire process in this route is slow, and comprises the secondary emotions.

Damasio proposed the somatic marker hypothesis to explain how emotions help people avoid repeating similar failures. The somatic maker is a hypothesized memory system that will fetch memories of unpleasant experiences which are relevant to a negative emotion. The invoked memory will then be used to remind the actor of potential dangers based on these previous experiences. For example, if

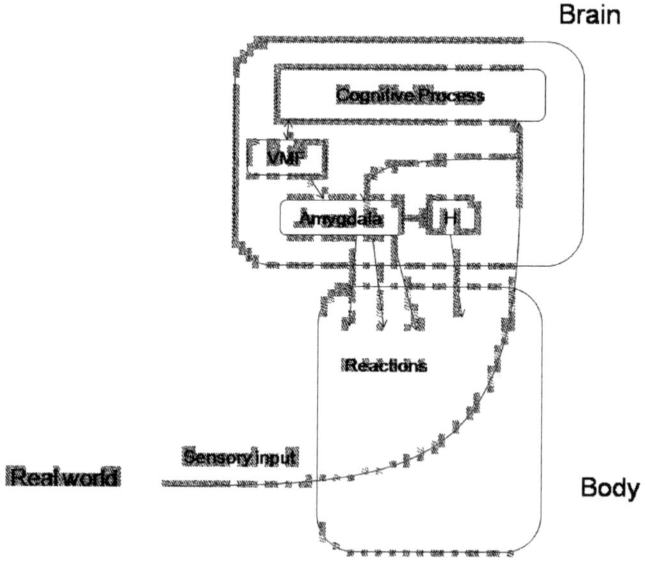

Fig. 7. Mechanism for secondary emotion [4]

one looses too much in a game, the emotion system might invoke an emotion of fear in the actor. It will in turn cause the actor's memory system to retrieve episodic memory concerning bad experiences in the past. The rational thought system may then use this information to foresee the potential outcomes in the current situation and react to stop the game for example. Thus, the emotional system helped the decision support system make a decision about an action that could lead to disastrous consequences.

Emotion is not only important as a mechanism for surviving by avoiding dangers based on experiences but is also a key used to achieve empathy in communication. The ability to simulate and infer the emotional status of other agents is critical to understand the communication partner, to interpret their reaction and to predict future events. The ability to create a proper emotional expressions is mandatory in order to derive a rapport with the partner.

Pickard suggested a computational architecture based on Damasio's model of primary and secondary emotions, as shown in Fig. 8 [6]. Pickard's model suggests that emotional states are accessible from both the inference and decision making subsystem (i.e., the upper-level module responsible for secondary emotions) and the pattern recognition and synthesis subsystem (i.e., the lower level module responsible for primary emotions).

Becker introduced dynamics to extend the PAD model[1] to cover secondary emotions [12]. The emotion (which is a short-lived valenced phenomenon) and the mood (which is a longer lasting valenced state) are combined to model the dynamical temporal behaviors of emotions (Fig. 9). Two independent spiral springs are simulated to model how the mood and valences are damped over time. Furthermore, a new variable is introduced to represent boredom. The range of boredom is [–1,0] (0: *zero* boredom, –1: boredom). The degree of boredom increases in the absence of stimuli. The value of boredom is reset to zero when the emotion component receives stimulus.

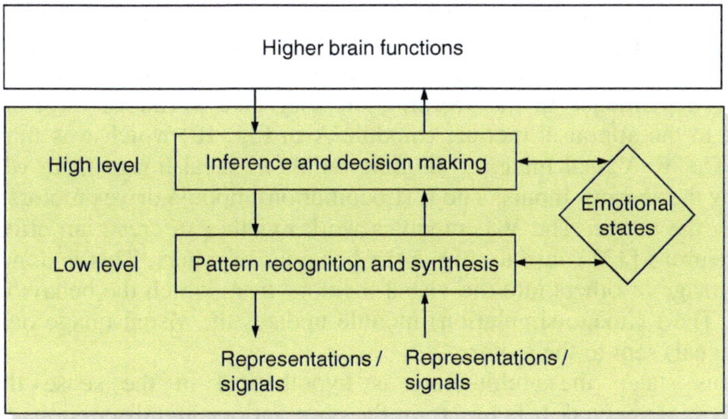

Fig. 8. Mechanism for primary and secondary emotions [6]

[1] See section 4 for introduction and more detailed discussions.

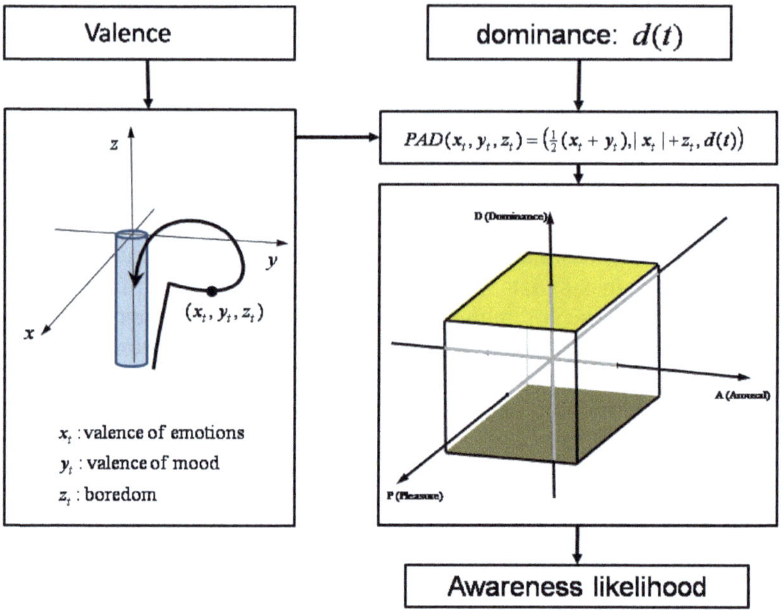

Fig. 9. Dynamical system model for primary and secondary emotion [12]

6 Emotional Intelligence in Image-Based thought Process

In addition to the distinctions between the primary and secondary emotions, authors in brain science have emphasized importance of images in the thought process. We consider an extension of the computational framework we introduced in Section 4.

The key idea is to characterize the working memory as a visually simulated world where images are intensively used and to introduce three "hardware" components for manipulating images. The modified architecture shown in Fig. 10 can handle visual images in the working memory. We introduce three modules in addition to the appraisal module (module A in Fig. 10) which was introduced in Fig. 3. The V (Visual imaging) module builds a visual image from what is provided by the sensory inputs. The L (Locomotion) module drives motors which are based on the image. The V-L modules work together to create an effect that the mirror neurons [13] cause to mimic the behavior of others. This is done by copying the image of others into the visual memory from which the behavior of self is created. The I (Image simulation) module updates the visual image based on the motor signals sent to the motors.

At this stage, the architecture is hypothetical in the sense that it has not been implemented. It is based on the observations and discussions reported in the literature. It only serves as a means for discussing emotional intelligence. It should be noted that scientists in neurophysiology have no consensus about how mental images are implemented in the brain. It might be a compound of

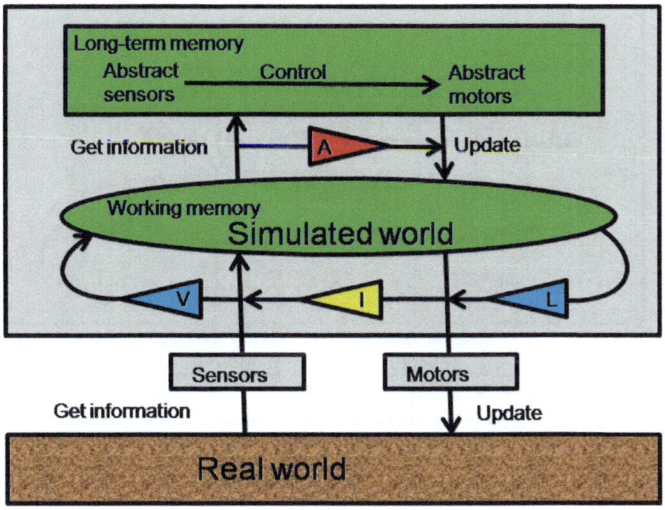

Fig. 10. Computational model for reasoning with mental images

low resolution facsimile images and features used to associate the visual image with the abstract symbolic representation. We do not make a commitment as to how the visual images are implemented.

The above framework works for a simple situation in which the agent is grasping an object (Fig. 11(1)). According to the hypothetical framework for the above, the V module will be invoked and a mental image created as a result (Fig. 11(2)). If the agent wants to reach and grasp the object, it will need a deliberate planning process to create a visual plan to achieve the aim (Fig. 11(3)). The I module might be invoked during the planning process in order to verify the achievement of the goal. This is done by simulating the body movement with the motor signal generated by the L module. The process might be carried out without moving the body. This mechanism may also be used to provide a visual feedback during the physical reaching-grasping stage (Fig. 11(4)).

The A module may add an emotional flavor to the process. For example, if the agent fails to grasp the object and drops it, the A module will create an alarm signal to alert the entire process so that the agent can invoke an immediate damage avoidance procedure. At the same time, cause-analysis-and-back-up processes may begin to analyze the cause and possible outcome of the event. It will cause the agent to form the secondary emotion and make a decision. Such an emotional process may significantly improve the quality of the reaction.

The idea of introducing a theater of mental images may be extended by introducing the theory of mind [14]. The key idea is to reuse the mental process to simulate the mental process of other agents by using the knowledge and reasoning mechanism of 'self'. This result can be realized by decoupling the mental process from self to reuse it to simulate the other's mental process in the mental world of self and to attribute the outcome.

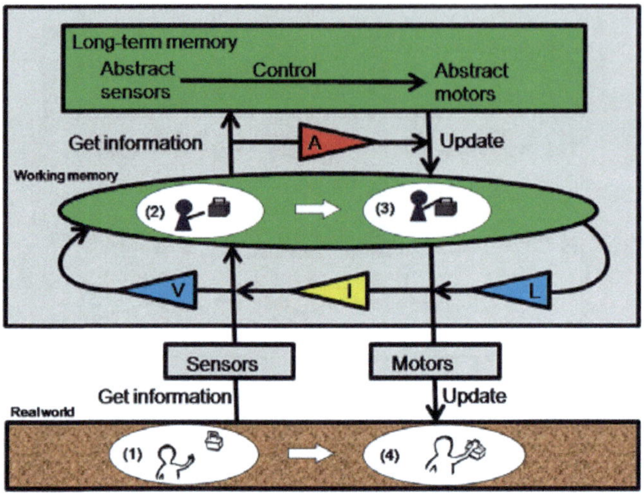

Fig. 11. Process of reasoning with mental images

Fig. 12 shows how the framework works for the wheelchair scenario [6], which approximately goes as follows:

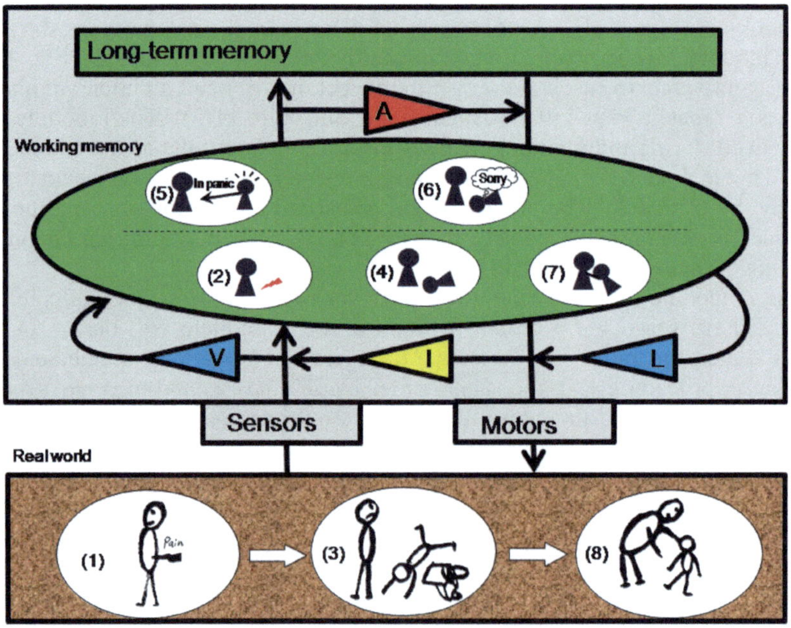

Fig. 12. Process of reasoning with mental images

> ... Suddenly, Rafe felt a piercing pain from a blow to his lower back. Rafe turned rapidly with an angry expression ... Rafe saw that he had been hit by Rebbeca, a person with hemiplegia whose wheelchair had gone out of control and caused her to crash into Rafe. ... Rafe's understanding that the cause of his pain was an uncontrollable event that had embarrassed Rebecca immediately changed his anger to sadness and sympathy. Though still in pain, his happy nature surfaced, and he began helping Rebecca recover from the accident. ([6], p. 23)

In this and previous example, the pain may trigger a primary emotion and the reasoning about the initial cause (Fig. 12(1)-(4)). We assume that the agent of Rafe can now use the mechanism for self (Rafe) to simulate the internal process of the other agent (Rebecca). It is now possible to interpret the incident and take into account the other agent's mental process. Thus, the panic of Rebecca before the collision or her mental status after the accident could be derived as a result of the simulation (Fig. 12(5), (6)). The embodiment and long-term memory of self are used to simulate the other's mental process. Simulation of the other's mental process would be harder if little information is shared with the other agent.

7 Emotional Intelligence in Communication

Emotional intelligence is a key used to make embodied conversational agents believable to the user. The media equation hypothesis [15] suggests that people treat socially the mediated humans in the same way as the real humans do at the unconscious level. In order for the Computers-Are-Social-Agent (CASA) paradigm [16] to be sustained in the human-agent interaction, the agents need to behave according to correct socio-emotional principles.

Emotional intelligence may also become relevant at a deeper level, in order to provide better services. Since humans are social and emotional, service provision systems in general should be able to sense the emotion of the user and be aware of the social context in order to guide the service provision to their goals. A good example is a piano-teaching computer illustrated in the opening chapter of [6]. This is a system that may nudge the students as long as it finds that they are motivated and still maintain interest. It may slow down and offer encouraging feedback when it detects signals that suggest the student is tired or frustrated. The principles underlying Pickard's piano-teaching computer apply to a wide variety of applications which include entertainment, navigation, and instruction, for example.

Becker shows a framework of embodied conversational agents in which emotion module is an integral part of a game playing agent MAX [12]. The dynamical system model shown in Fig. 9 is used as an emotion module. This is connected to a dialogue manager and consists of two control layers corresponding to primary and secondary emotion. The dialogue manager provides the valence and dominance values for the emotion module. In turn it calculates the awareness likelihood value for the dialogue manager in order to filter the emotions.

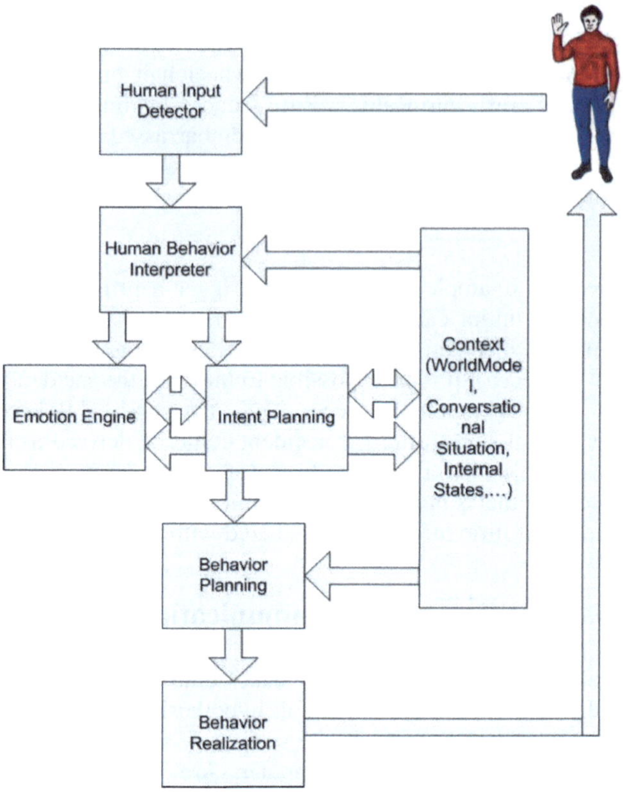

Fig. 13. The architecture of a customer service agent [18]

At the finer grain level, Lance et al show a method of creating the emotional motion of the head and body that occurs with eye gaze [17]. Basically, the method modulates the movement of the head and body according to the emotions.

Fig. 13 shows the architecture of a customer-service agent system designed to interactively give instructions when using complex instruments [18]. An emotion engine is introduced to provide a personal trait to support emotionally adequate interactions with the user. The behavior of the agent results from the interactions between the emotion engine and the planning modules.

References

1. Goleman, D.: Emotional Intelligence: Why It Can Matter More Than IQ, Bantam (1995)
2. Goleman, D.: Social Intelligence: The New Science of Human Relationships. Bantam, NY (2006)
3. Bechara, A., Damasio, H., Damasio, A.R., Lee, G.P.: Different contributions of the human amygdala and ventromedial prefrontal cortex to decision-making. The Journal of Neuroscience 19(13), 5473–5481 (1999)

4. Damasio, A.R.: Descartes' Error: Emotion, Reason, and the Human Brain. Gosset/Putnam Press (1994)
5. Ortony, A., Clore, G.L., Collins, A.: The Cognitive Structure of Emotions. Cambridge University Press, Cambridge (1988)
6. Pickard, W.R.: Affective Computing. The MIT Press, Cambridge (1997)
7. Breitenberg, V.: Vehicles: Experiments in Synthetic Psychology. MIT Press, Cambridge (1984)
8. Maes, P., Blumberg, B., Darrell, T., Pentland, A., Wexelblat, A.: Modeling Interactive Agents in ALIVE. IJCAI 1995, 2073–2074 (1995)
9. Bates, J.: The role of emotion in believable agents. Communications of ACM 37(7), 122–125 (1994)
10. Russell, J.A., Mehrabian, A.: Evidence for a three-factor theory of emotions. Journal of Research in Personality 11, 273–294 (1977)
11. Mehrabian, A.: Pleasure-arousal-dominance: a general framework for describing and measuring individual differences in temperament. Current Psychology: Developmental, Learning, Personality, Social 14(4), 261–292 (1996)
12. Becker, C., Kopp, S., Wachsmuth, I.: Why emotions should be integrated into conversational agents. In: Nishida, T. (ed.) Conversational Informatics: An Engineering Approach, pp. 49–68. Wiley, Chichester (2007)
13. Rizzolatti, G., Sinigaglia, C.: Mirrors in the Brain – How our minds share actions and emotions, translated by Anderson, F. Oxford University Press, Oxford (2008)
14. Baron-Cohen, S., Leslie, A.M., Frith, U.: Does the autistic child have a 'theory of mind'? Cognition, vol. 21, pp. 37–46 (1985)
15. Reeves, B., Nass, C.: The Media Equation: How People Treat Computers, Television, and New Media Like Real People and Places. Cambridge University Press, Cambridge (1996)
16. Nass, C., Steuer, J., Tauber, E.R.: Computers are social actors. In: Proceeding of the CHI Conference, pp. 72–77 (1994)
17. Lance, B., Marsella, S.C.: Emotionally Expressive Head and Body Movement During Gaze Shifts. In: Pelachaud, C., Martin, J.-C., André, E., Chollet, G., Karpouzis, K., Pelé, D. (eds.) IVA 2007. LNCS (LNAI), vol. 4722, pp. 72–85. Springer, Heidelberg (2007)
18. Hacker, B.A., Wankerl, T., Kiselev, A., Huang, H.H., Merckel, L., Okada, S., Schlichter, J., Abdikeev, N., Nishida, T.: Incorporating Intentional and Emotional Behaviors into a Virtual Human for Better Customer-Engineer-Interaction. In: Proc. 10th International Conference on Telecommunications (ConTEL 2009), Zagreb, Croatia, June 8-10, 2009, pp. 163–170. The University of Zagreb (2009)

Chapter 2
Automatic Affective Dynamics: An Activation–Habituation Model of Affective Assimilation and Contrast

Kari R. Irwin, David E. Huber, and Piotr Winkielman

Department of Psychology
9500 Gilman Drive
University of California, San Diego
La Jolla, CA 92093-0109
krirwin@ucsd.edu

Abstract. Our modeling approach seeks to better understand the computational dynamics of the affective and cognitive systems. One experimental phenomenon open to such dynamical analysis is "affective priming" -- or the influence of a prior stimulus on subsequent affective processing. In this type of procedure, 'assimilation priming' refers to when the response to the target is biased in favor of the prime, such as with brief or minimally attended primes. However, following long durations or highly attended primes, the response to the target is often biased against the prime, which is termed 'contrast priming'. We present a neural dynamics model of affective priming in which this transition from assimilation to contrast occurs automatically as a result of habituation. Unlike response strategies, this transition is predicted to rise and fall in a gradual nonlinear manner as a function of prime duration. We confirmed this prediction with a speeded affect judgment task that manipulated the exposure duration of valenced images.

1 Introduction

> "In my own opinion, affective habituation is a phenomenon of the same order as sensory adaptation, and results always and only in indifference. Even if Ebbinghaus is correct, and quality passes into opposite quality, we have a sensory analogy in the case of vision: adaptation to yellow means blue-sightedness, local adaptation to green means a purple after-image."
>
> Edward Titchener.

As proposed by Titchener in 1908, affective responses to emotionally valenced stimuli can produce habituation similar in nature to that of perceptual habituation. If this account is correct, then affect is a response that naturally and automatically rises and falls when a valenced stimulus is presented for an extended time period.

T. Nishida et al. (Eds.): Modeling Machine Emotions for Realizing Intelligence, SIST 1, pp. 17–34.
springerlink.com

By analogy, consider the response of the visual system when first stepping into bright sunlight. Initially everything is blindingly white, but habituation normalizes brightness levels and the outside world comes into view. Further habituation occurs over the course of seconds or minutes, which allows the visual system to adjust to the ambient light levels until the bright sunlight becomes comfortable. However, upon returning indoors, the opposite occurs. Initially the indoor world is too dark to see until habituation again normalizes for the ambient light levels. Indeed, it is believed that pirates on the high seas often wore a single eye patch to sidestep the slowness of habituation (Van de Water 1942). By wearing a patch while on deck, one eye was always habituated to darkness even though the other was habituated to sunlight. This way, pirates could see during brief trips to the nearly complete darkness below deck simply by flipping up their eye patch.

As noted by Titchener, perceptual habituation does not literally produce an opposite response, but instead it tempers the response to the stimulus at hand. Because habituation takes time to dissipate, this produces an apparent contrast effect when presented with a new stimulus (e.g., a return to the indoors) following extended exposure to an extreme stimulus (e.g., bright outdoor sunlight). In this chapter, we implement Titchener's suggestion by developing a dynamic neural network of habituation as applied to affective responses. Unlike the presented model, previous theories of affective priming proposed that controlled decision processes underlie contrast effects. However, automatic habituation was previously demonstrated to provide an accurate account in the domain of priming with word identification (Huber and O'Reilly 2003).

According to this account, habituation serves to normalize to the ongoing state of the world. So, in regards to affect processing, positive affect soon habituates in the presence of a stimulus that elicits a smile (e.g., an image of a puppy). By normalizing in this way, the affect system is made more sensitive to changes. Thus, exposure to a new stimulus (e.g., an image of torture), is easily detected as notably different. In this manner, normalization through habituation serves to illuminate variety in the landscape of affective events.

Although the presented affective habituation model cannot falsify strategic accounts of affective priming, we appeal to parsimony considering that these neural dynamics are known to exist, and considering that these dynamics have previously explained perceptual priming. After presenting the model equations and structure, we report model predictions as a function of prime duration. Finally, we present a new experiment with valenced images for primes and targets that confirmed these predictions.

2 The Effects of Priming: Unwarranted Influence

In psychology, the term "priming" generally refers to the overall influence an earlier stimulus has on later processes. This influence is considered to be fairly automatic in nature (occurring below the level of conscious awareness), and under some theories is thought to be driven by the overlapping perceptual or conceptual features present in earlier and later stimuli.

2.1 Perceptual and Semantic Priming

Perceptual priming is based on the physical form of the stimulus. For example, Huber, Shiffrin, Lyle, and Ruys (2001) used a two alternative forced choice procedure (2 AFC) to indirectly measure the effects of priming visual features on later perceptions. In their paradigm, a prime word was presented, followed by a briefly flashed target word (i.e., around 50 ms), then a mask, and finally a choice between two words. With this presentation sequence for repetition priming (i.e., the prime and target word are identical), it was observed that primes viewed passively for a short duration (i.e., 500 ms) resulted in an "assimilation effect" (i.e., better performance for similar targets as compared to trials with dissimilar targets). However, following primes viewed passively for a longer duration (i.e., 2,000 ms), a full reversal of these effects was observed (i.e., a 'contrast effect'). This suggests that when the prime duration is increased to certain durations, later processing can become impaired for similar targets as compared to dissimilar targets.

Unlike perceptual priming, however, in which assimilation and contrast effects are dependent on the physical form of a stimulus, conceptual priming is driven by semantic or associative features of the stimuli. Even so, the types of priming effects observed in both perceptual and conceptual priming are quite similar. For example, Meyer and Schvaneveldt (1971) showed a similar benefit based on shared prime-target meaning, such as with semantic or associative priming. In this experiment, subjects were either to press one key if both of two simultaneously presented visual letter strings were English words or to press another key otherwise. Interestingly, subjects were again faster and more accurate in responding to displays containing two semantically related words (e.g., 'bread' and 'butter') as compared to displays with two unrelated words (e.g., 'doctor' and 'butter'). The claim here is that like the automatic activation of visual features, semantic features are automatically elicited as well. In both cases, the residual activation of visual or semantic information from the first stimulus affects processing of subsequently presented items.

Priming experiments that manipulate the semantic relatedness of stimuli are used to study the representations used in language. By activating a mental construct with the mere presentation of a word, researchers are able to explore the structure by which words are stored in memory, as well as the process by which they are automatically activated and processed. The same is true for affective processing. Researchers can utilize a variety of priming procedures, presenting emotionally valenced words or images as stimuli, in order to better understand the process by which affective responses are generated and the degree to which affective states share a conceptual representation.

2.2 Affective Priming

In using traditional priming procedures with affective words or images as stimuli, the storage and retrieval of affective constructs can be targeted. And strikingly, it seems as though there are many similarities in how both neutral items (e.g., words) and affective items (e.g., emotional images) are integrated, stored, and retrieved.

A study by Fazio, Sanbonmatsu, Powell, and Kardes (1986) was one of the first to extend from previous semantic priming studies. It documented that, like semantic constructs, which are shown to be automatically activated, evaluative representations are similarly activated in an automatic way. In this study, participants were asked to indicate whether a target word was positive or negative in valence as quickly and accurately as possible. What Fazio et al. (1986) observed was that participants were often faster to make their response when the prime and target were evaluatively congruent (i.e., both positive or both negative) as opposed to when they were evaluatively incongruent. In other words, word pairs sharing only valence (i.e., an intrinsic positive or negative emotional value) produced the same type of assimilation effect as perceptually or conceptually similar words. This suggests a general effect of overlapping prime-target features, regardless of whether or not those features are valenced, and whether they are concrete or abstract in nature.

2.3 Assimilation and Contrast Effects in Priming

The field is more or less in agreement that assimilation effects in priming occur automatically. That is, the facilitated responding for similar items seen in priming occurs below the level of consciousness and as a result of the automatically activated representations (affective or not), which aid in later processing. The underlying nature of contrast priming, however, is still heavily debated.

2.3.1 Strategic Account of Contrast Effects

Traditionally, contrast in priming has been assumed to be strategic in nature. That is, while assimilation priming is likely automatic, contrast is thought to be due to a consciously implemented adjustment to the prime's influence. The implementation of such a compensatory strategic process is most likely when individuals are aware of the prime, as well as motivated and capable of correcting for the prime's unwarranted influence. These factors were confirmed in a number of priming studies. For example, the role of conscious awareness in producing contrast effects is demonstrated in a study where supraliminally presented primes were more likely to produce contrast effects than subliminal presentations of the same image (Murphy and Zajonc 1993). Additionally, Lombardi, Higgins, and Bargh (1987) observed that contrast effects were also more apparent when the primes were explicitly remembered versus primes that were not remembered. And finally, Martin, Seta, and Crelia (1990) suggested that sufficient available cognitive resources were necessary for contrast priming, given that such effects were less likely when participants were cognitively distracted.

In sum, there are numerous studies that have supported a strategic account of contrast priming, and importantly, it is not our intent to refute the role of conscious processes in mediating priming effects under some conditions. However, it is our current goal to support the relatively new stance that contrast priming could result, in part, from automatic processes as well.

2.3.2 Automatic Account of Contrast Effects

In support of the claim that an automatic mechanism may play a mediating role in affective priming, Glaser and Banaji (1999) demonstrated a full cross-over from assimilation to contrast priming simply as a function of the extremity of the prime. Participants in this study were asked to make an objective valence judgment about a series of target words (i.e., whether the word was positive or negative). In doing so, they showed a traditional assimilation effect only when both the prime and target were moderately valenced (e.g., 'butterfly' or 'fire'). Instead, if the prime word was very extreme in nature (e.g., 'love' or 'vomit'), then the effect fully reversed to contrast priming. After replicating these effects, Dijksterhuis and Smith (2002), further observed that repeated subliminal exposure to an extreme stimulus can cause it to behave as a moderately extreme stimulus. That is, the extreme stimuli, which would typically produce a contrast effect following a single exposure, instead resulted in an assimilation effect following multiple subliminal exposures.

The cross-over transitions documented by Glaser and Banaji (1999) and Dijksterhuis and Smith (2002) support the role of an automatic mechanism, because the implementation of a strategy is highly unlikely. A strategic interpretation would require that on a trial by trial basis, subjects either adopted or did not adopt a contrast strategy depending on the particular prime that was presented. Thus, in order to provide a more parsimonious and over-arching account of priming results, we propose an automatic underlying mechanism. But more specifically, we propose that the automatic mechanism at play is related to the fundamental process of habituation.

2.3.3 Automatic Contrast: An Habituation Account

In general, habituation refers to the automatic change in responsiveness of the sensory system to a constant stimulus over time. For example, perceptual phenomena like afterimages and the motion aftereffect are the result of neural adaptation of visual receptors followed by an increased sensitivity to changes in the visual field (Barlow and Hill 1963). However, in terms of affective processing, an habituation account proposes that adaptation causes the overall intensity of an affective response (e.g., the response derived from a valenced word), to gradually fade over time to a diminished level. This adaptation allows the affective system to rapidly adapt to its current state, becoming habituated to the good or the bad, and therefore more sensitive to relative changes in emotional status.

Habituation theory also postulates that it is the automatic adaptation to an affective response that produces contrast effects in priming. Suppose a case where the prime and target share a neural representation (e.g., positive valence). At short prime durations the prime is only partially processed at the time the target is presented, and the residual activation will aid in its processing (resulting in assimilation priming). However, at longer durations, the prime is fully processed, and the affective response is habituated, at the time the target is presented. Thus, the affective system is highly sensitive to changes in valence (e.g., an incongruent target), yet sluggish to reactivate the shared construct for the congruent target (resulting in contrast priming).

More generally, this theory of habituation suggests that both assimilation and contrast effects result from automatically changing levels of neural activation. Furthermore, the theory suggests that these dynamic effects of priming are linked to the degree to which a prime stimulus has been processed. Stapel, Koomen, and Ruys (2002) observed priming effects consistent with this idea. With only a manipulation of prime duration, they demonstrated a transition from assimilation to contrast as prime exposure increased from 30ms to 100ms. Our account of habituation predicts this sort of transition, since at short durations prime activation is still accumulating, thus related targets will receive some benefit. But, at longer prime durations, the prime is fully processed and habituated, making related targets more difficult to process.

These manipulations directly affect the extent to which the prime can be processed. But importantly, the results also begin to (1) provide a time course for how long affective stimuli take to process, (2) quantify the direction and magnitude of the prime's influence on later affective processes, and (3) allow postulation about underlying mechanisms that could potentially explain such non-intuitive patterns of results.

3 A Neural Habituation Account of Priming Effects

As mentioned previously, assimilation and contrast effects are evident in both perceptual and affective priming experiments, which provide reason to postulate a single automatic underlying mechanism to account for these effects. Even though these very different types of priming tasks each explore different types of processing, the results across experiments are characterized by strikingly similar effects. For example, the automatic activation of a construct from the mere presentation of a stimulus is not specific to affect, but rather functions similarly for semantic, orthographic, and other visual features. Additionally, priming effects in perceptual identification tasks exhibit the same cross-over transition from assimilation to contrast as a function of increasing prime duration, as is seen in affective priming experiments (Huber et al. 2001; Stapel et al. 2002).

The similarities across visual, semantic, and affective priming, lead us to propose that a single mechanism of activation and habituation underlies many types of processing more generally. Under this supposition, we then seek to extend an existing computational model of priming instead of developing a theory and model from scratch to account for a specific effect in affective processing.

The neural network model nROUSE (Huber and O'Reilly 2003) used residual activation and habituation to explain the priming effects observed in a word identification task. This model adequately accounts for observed perceptual priming effects with automatic mechanisms of perception and cognition. It assumes that decisions for target words are based on features (e.g., visual, orthographic, semantic), and, that both the extent to which the prime is processed and the degree to which the prime and target share features, will result in beneficial or detrimental effects on the processing efficiency of the target.

The nROUSE model of neural habituation was developed and tested within the basic processes of perception, priming, and memory. And, while it may seem a stretch to generalize such a model to more abstract affective processes, the dynamics of habituation and persistence exist throughout the brain. Thus, these dynamics should exist for all kinds of processing, including affect. Furthermore, this single account will demonstrate that the paralleled effects across domains are due to a common underlying mechanism and provide a parsimonious explanation for the observed patterns of priming effects.

The Huber and O'Reilly (2003) model in particular provides an account of habituation. It is grounded within fundamental neural processes, and, activation within the model's artificial neurons follows the same pattern of true neural activation as was observed by Tsodyks and Markram (1997). Both the observed electrophysiological recordings and the simulated firing rates show an initial accumulation of activation upon exposure to a stimulus, which peaks after some interval, and is followed by accommodation (or neural habituation). On this account, this accommodation occurs in order to reduce excess activation, conserve neural resources, and minimize any residual activation (i.e. prime persistence) after the removal of a stimulus.

There are several neural mechanisms that can explain this sort of 'activity-dependent accommodation', such as neurotransmitter depletion; however, regardless of the specific mechanism, the resultant neural behavior is well described as synaptic depression (Nelson, Varela, Sen, and Abbott 1997; Tsodyks and Markram 1997). Thus, we focus on synaptic depression (i.e., resource depletion), and use the dynamics of this process to derive a general neural network model.

Synaptic depression occurs as resources in the synapse (e.g., neurotransmitters, calcium, etc.) become depleted due to the activation of the pre-synaptic cell. Then, because of this depletion, the transferred activation between the two cells, or the overall effect of each action potential on the receiver neuron, is also greatly reduced. The dynamics of this process have been studied with electrophysiological recordings of both the pre- and post- synaptic neurons, and these recordings have helped to illustrate the magnitude and time scale of these effects. An example of such electrophysiological recordings is displayed in Figure 1 (Tsodyks and Markram 1997), which illustrates how, even with a constant activation applied directly to the pre-synaptic neuron, the post-synaptic neuron has an initial depolarization, followed by a rapid decrease to an asymptotic level after about 200ms.

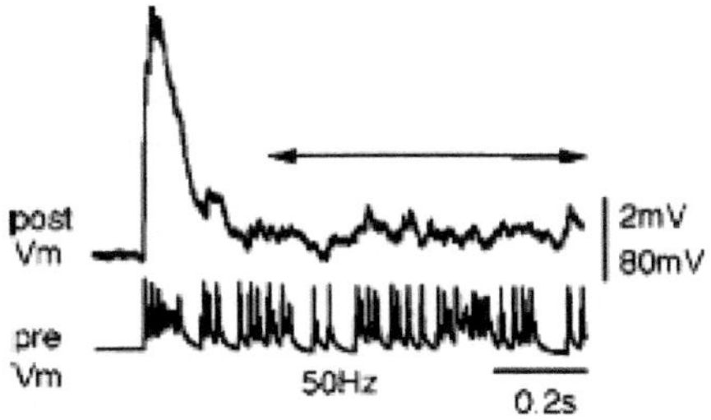

Fig. 1. Electrophysiological recordings obtained by Tsodyks and Markram (1997) with patch clamping of two connected neurons. This captures the processes of neural activation and subsequent habituation as dictated by the process of synaptic depression.

4 A Neural Network Model of Affective Priming

Using the previously described principles of neural activation and habituation, we next propose a neural network model that incorporates the dynamics of habituation and resource depletion to account for the effects observed in affective priming tasks. Previous research regarding synaptic depression has already observed several perceptual and behavioral correlates. For example, Pantic, Torres, and Kappen (2002) demonstrated that synaptic depression allows for rapid switching between memory states; Chance, Nelson, and Abbott (1998) showed that synaptic depression in the primary visual cortex allows for increased detection of moving/changing stimuli; and, Gotts and Plaut (2002) showed that damage to the synaptic depression process is a possible explanation for neuropsychological deficits in semantic processing. Because the low level process of synaptic depression correlates with several aspects of higher level functioning, we suspect that depletion of resources at the neural level may play a similar mediating role in affective processing.

4.1 Model Structure

Within the proposed model, each artificial neuron (i.e., node) individually incorporates characteristics of synaptic depression. It is the inclusion of this mechanism that produces more realistic patterns of activation for each node, in which output activations are similar to the transient dynamics observed in single cell recordings (see Figure 1; Tsodyks and Markram 1997). Each node itself is then placed within a cascading hierarchical network through which activation can travel, as seen in Figure 2 below.

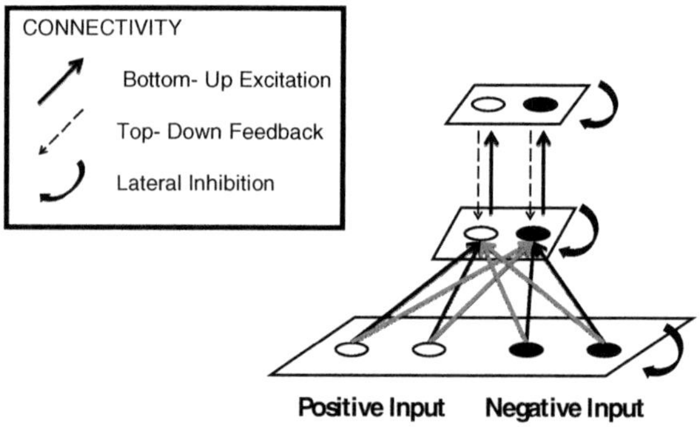

Fig. 2. The hierarchical structure for the proposed habituation model of affective processing. From bottom to top, the levels within the network represent progressively more complex features of the stimuli. There are two pathways within the network representing positive and negative valence. However, the projections from the lowest to middle layers are proportionally distributed with 85% traveling to the node of the same valence and 15% traveling to that of the opposite valence. Connections between levels in both directions are excitatory. Connections within each level are inhibitory. See text for further discussion and details.

We do not have results that constrain the exact nature of each level within the neural network in the current application of the model. So, given the lack of necessity at this point, there are no strong claims about what the hierarchical layers specifically represent or which brain regions might be involved. But rather, the structure represents the general integration process of information, in which simple visual features are combined to form a more complex representation from which an evaluative response can be made.

The model does require, however, that the integration process occurs within three distinct stages. These layers are required in order to produce both the assimilation and contrast effects like those observed in perceptual and affective priming. Assimilation effects (i.e., when the target response is biased in favor of the prime) occur as a result of source confusion, which in the model is due to converging activation between the bottom and middle layers. On the other hand, contrast effects (i.e., when the response to the target is biased in the opposite direction of the prime) emerge as a result of the habituation association after the point of convergence (i.e., the connection between the middle and highest levels). Additionally, lateral inhibition is required within each layer of the network to dampen excess activation and prevent multiple representations from being equally active at a given time.

The hierarchical network also defines two separate network streams through which activation can travel, representing both positive and negative valence. Each of these streams is made up of two nodes in the lowest visual layer, which allows for visual input from both the prime and target in the case that they are of the same valence. Then, each node in the lowest layer projects to both the positive and negative node in the middle layer, with the projected activation proportionally distributed and preferential to the node of the corresponding valence. This distributed representation is proposed in the model because valenced stimuli are often highly complex. An emotionally valenced image might easily contain both positive and negative elements when these elements are considered in isolation, even when the sum of those elements is clearly one or the other. For instance, take the image of a gun pointed at a child. The scene as a whole is obviously negative; however, the mere presence of a child's face may still activate some positive feelings on some level. This type of stimulus complexity is true for valenced visual scenes encountered in real life, as well as those used in the later presented simulation study and experiment. Thus, it seems necessary that any model of affective processing incorporate this element of proportionally distributed activation of multiple emotional states.

The balance and interaction between opponent forces of emotions (such as positive versus negative), is in line with applications of opponent process theory that are decades old. This theory was initially described by Hurvich and Jameson as a general model of neurological opponent processing for phenomena such as color vision (Hurvich & Jameson 1957). However, this model was later expanded to explain emotional behavior (Solomon & Corbit 1974). The extended model asserts that opposing emotions are paired, and that when one emotion in a pair is experienced, the other is suppressed. It also predicts an opponent reaction in the opposite direction after the primary affective response has been habituated. In effect, the theory claims that emotions modulate around a point of neutrality when stimulated, or that opponent forces can cancel each other out. This theory has been supported in studies in which subjects were asked to report feelings of different emotions (e.g., fear and pleasure) while engaging in various activities (e.g. skydiving) (Solomon and Corbit 1974). However, the currently presented affective habituation model proposes a neural mechanism by which opponent emotions may interact. With a specific network structure and dynamics of activation, this model offers falsifiable predictions and measures with which to quantify the interacting effects of opposing emotional states.

4.2 Activation Dynamics

Each individual node within the proposed network functions to capture aspects of temporal integration of information through dynamic activation updates. As initially described by Huber and O'Reilly (2003), these updates are expressed as a probability of spiking over some time interval. This expression is used as opposed to a discrete number of spikes, because it allows for each artificial neuron to be viewed as a general representation of any neuron functioning under similar inputs and outputs, simplifying the biological processes. The simulated membrane potential, v, in each of the nodes (neuron i in layer n) is updated as a function of weighted excitatory inputs, lateral inhibition, and leak currents as a change in potential per unit of time. The general form of this membrane potential update is:

$$\frac{\Delta v_i^n(t)}{S_n} = (1 - v_i^n)\left\{\sum_{\forall j} w_{ij} o_j^{n-1} + F \sum_{\forall k} w_{ik} o_k^{n+1}\right\} - v_i^n \left\{L + I \sum_{\forall l} o_l^n\right\} \quad (1)$$

As can be seen in Equation 1, the weighted bottom-up and top-down excitatory inputs combine to drive the membrane potential towards the excitatory reversal potential of 1.0 (i.e., the first term in the curly brackets, multiplied by 1-v). The weights for the bottom-up connections are given a default strength of 1.0, whereas the weights for the top-down feedback are given the strength value F ($0 < F < 1$). In opposition to this excitatory input, lateral inhibition (i.e. the summation over the l unites within level n) is given a strength of I, and combines with a fixed term for leak, L. Importantly, it is the slow leak of activation that occurs within the membrane that results in residual activation (or persistence of the prime). Jointly, the facilitory and inhibitory inputs drive the membrane potential towards the reversal potential of 0 (i.e., the second term in curly brackets, multiplied by $-v$).

Importantly, not every layer within the structure is restricted to integrate information at the same speed. Instead, the overall differences between the layers are accommodated for with the parameter Sn. This free parameter serves to scale each of the update calculations for each of the three layers, regulating the speed at which information is integrated over time. The larger the value of Sn, the faster information is integrated, thus changing the membrane potential more quickly. Conversely, smaller values of the Sn parameter result in the layer integrating information more slowly.

Equation 2 below specifies the update calculation for synaptic output. But, unlike traditional artificial neural networks whose activation value varies only with time, this model represents the synaptic output as a product of two time varying parameters. These dynamic parameters being: (1) the amount by which the membrane potential exceeds some threshold value (Θ), and (2) a factor, a, representing available resources for that particular connection. In other words, synaptic output is the product of the on-average firing rate (v-Θ) and the effect of each action potential given the available resources (a). This dynamic calculation thus captures the process of habituation due to recent activity, and predicts how systems of neurons will be affected by this resource depletion (Abbott et al. 1997; Tsodyks & Markram 1997).

$$o = (v - \Theta)a \tag{2}$$

Equation 3 provides the update calculation for synaptic resources available in a given connection (i.e. neurotransmitters, calcium, etc), assuming mathematical constants for the parameters D (depletion) and R (resources). The process of synaptic depression is represented in the equation by two parts. First, resource depletion is driven directly by the recent firing of the pre-synaptic neuron, and secondly, the extent to which those resources have been depleted is what drives the rate of recovery.

$$\frac{\Delta a_i^n(t)}{S_n} = R(1 - a_i^n) - Do_i^n \tag{3}$$

4.3 Mapping the Model to Responding

All three of the equations above, work in conjunction with one another to continuously update neural activation and resources available in the synapse. And together, they are able to capture patterns of responding. The model assumes that visual input travels primarily in a bottom-up fashion, until it reaches layer three, at which point performance that is based on affective evaluation is achieved through the fluency responses (i.e., time-to-peak) from the output of the highest level. Within these highest nodes, target activation accumulates, but quickly reaches a peak level. It is at this point that habituation becomes strong enough to limit the accrual of additional activation. The participant's categorical response (e.g., "positive" or "negative") is assumed to be the response whose corresponding activation reaches its maximum first following the presentation of the target item (i.e., a horse race). The actual time to respond is assumed to be related to this 'time-to-peak' of the winner. Thus, this measure is used as a relative indicator of response time between conditions, as it directly corresponds to the true reaction time.

5 Testing the Model: Experimental Procedure Used for Simulation Fit and Behavioral Study

In order to test the generality of the proposed habituation theory and dynamic model, as applied to affective processing, we next present both behavioral and simulated data for an experiment that parametrically varied prime duration. Behavioral data were collected for two reasons: (1) to observe the effect of this manipulation on human behavior, and (2) to serve as a test for whether the model was able to account for these effects. Qualitatively, the theory predicts that as a function of increasing prime duration, assimilation effects will gradually flip to contrast effects (rather than an abrupt flip, such as with a strategy). However, this particular model implementation of the theory specifies a particular functional form to this transition that may, or may not, be compatible with empirical results.

The experimental procedure (see Figure 3) involved the presentation of a single prime image followed by a single target image. Each image was specifically selected according to normative ratings of positive and negative valence (Lang, Bradley, and Cuthbert 1999). Every experimental trial began with a fixation stimulus ("+"), followed by a prime picture, which was displayed for 500, 1000, 1500, 2000, or 2500 ms. Following that exposure, a target image then replaced the prime and remained on the screen until participants indicated the valence of the image with a response key. The influence of the prime on a given target response was measured by reaction time, and the time course of the observed priming effects was made possible by the incremental prime durations that were used.

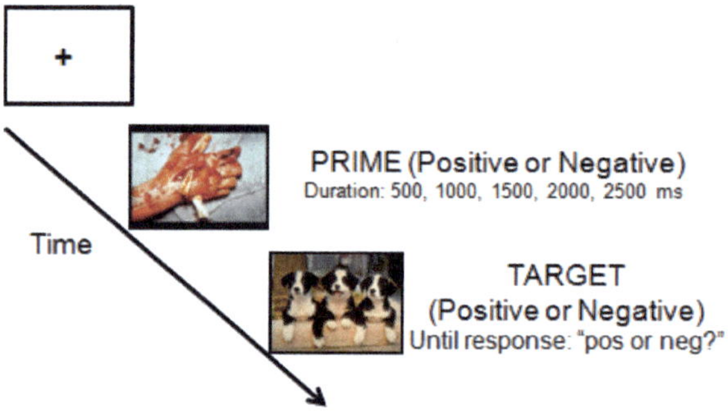

Fig. 3. Affective priming procedure used in the subsequent simulation study and behavioral experiment (Irwin, Huber, & Winkielman, in preparation)

A transition from assimilation to contrast was predicted because of the previously discussed rise and fall of neural activation upon exposure to any stimulus. When a prime is presented for only a brief duration, the level of activation is quite high at the time the target appears, thus assimilation is observed due to source confusion. However, when a prime is presented for a longer period, the meaning of the prime is more fully processed, which results in habituation to the features of the prime. This habituation makes the affective system sluggish to reactivate the shared representation (as with similar target items), yet quick to detect changes in status (as with dissimilar target items). Therefore, with increasingly long prime durations, assimilation effects are predicted to appear and disappear, and then fully reverse to contrast effects in a gradual, non-linear manner.

Additionally, as prime duration is increased to an even longer duration, it is predicted that all priming effects disappear completely. This final disappearance is a unique prediction of using a distributed coding scheme, and would not be the case if positive and negative processing occurred independently of one another. This distributed representation was used because of the complexity of the valenced stimuli, which are comprised of a number of elements ranging in affective value. Because of the this assumption, when prime exposure is sufficiently long, both the dominant response implied by an image, as well as the lesser response implied by an image, reach a level of full habituation. At that point, there is very little difference between the two responses, and so the contrast priming is reduced.

5.1 Behavioral Data: Prime Duration

Using the experimental procedure described in Figure 3, data was collected from 63 subjects, each run through 1 session comprised of 80 trials. Each individual trial included the random presentation of a single prime image for a variable duration (500, 1000, 1500, 2000, 2500 ms) and a single target image, which remained on the screen until the subject indicated its valence. In half of the trials the prime and target were congruent in valence (i.e., positive – positive or negative – negative), while in the other half, the prime and target were incongruent in valence (i.e., positive – negative or negative – positive). The indirect measure of the prime's influence was the time it took for the subject to indicate their response through a key press. The median reaction times for congruent and incongruent trials are displayed in Figure 4, as well as the relative difference between the two (incongruent trials minus congruent trials). Both plots show the observed change in target response time as a function of both prime duration and valence congruency.

The behavioral data do show the predicted non-monotonic effect. Importantly, this pattern of effects seems to confirm an automatic mechanism, as a strategic account would necessitate that subjects adopt a different strategy at each prime duration.

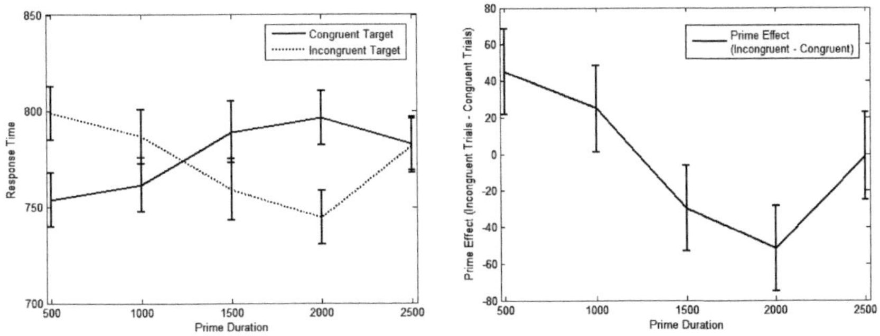

Fig. 4. Behavioral results from Irwin et al. (in preparation), which show the median correct reaction time results as a function of prime–target valence congruency and prime duration. The plot on the left shows the reaction times for congruent trials and incongruent trials separately, while the plot on the right shows the relative difference between the two priming conditions (i.e. the 'prime effect'). Error bars denote plus or minus one standard error of the mean.

5.2 Simulated Data: Prime Duration

We now present a test of the proposed habituation model, in which we assess whether or not it can account for the reported empirical findings. The simulation was run in time steps of one millisecond, where input was sent directly to the appropriate node within the lowest layer. Input was set to zero in general, but changed to one for the time steps when a visual object was presented that was the preferred input for that node. The values of the fixed parameters were left the same as those initially proposed by Huber and O'Reilly (2003). These parameters have worked well for both word and face priming and represent general cortical dynamics, so they should be applicable for affective processes as well (Huber & O'Reilly 2003; Rieth and Huber 2005). The values were: inhibition, $I = 0.3$, threshold, $\Theta = 0.15$, leak, $L = 0.15$, depression, $D = 0.324$, recovery, $R = 0.022$ and first layer's rate of integration, $S1 = 0.054$.

While these default parameters represent dynamics that have been studied and previously confirmed, the rates of integration for the middle and top layers ($S2$ and $S3$, respectively) represent novel affective elements. Thus, these parameters were adjusted to best fit the observed data. The best fitting values for these parameters were: the rate of integration for the middle level, $S2 = 0.005$, and for the highest level, $S3 = 0.03$.

The parameter specifying the proportion of activation sent to the dominant response versus the alternative response was also adjusted to best fit the observed data. Some distributed representation for the images was predicted due to their complexity (i.e. that the pictures likely evoked both a positive and negative response to some degree), however, the exact proportion of this distribution was unknown and therefore explored. For this parameter, the best fitting value was with 85% of activation being sent to the preferential node (i.e., to the 'correct' overall valence), leaving the additional 15% sent to the non-preferential node (i.e., the 'incorrect' overall valence).

Using the proposed structure and parameter values, the model captured the qualitative change in priming effects that were observed empirically. The similarities can be observed by comparing Figure 4 (i.e., the empirical data) with Figure 5 (i.e., the simulated data), which show that the observed and simulated reaction times follow a similar course as a function of prime duration and valence congruency. In moving from left to right on the graph, both figures show an initial appearance of assimilation effects at the shortest prime durations (i.e. congruent trials are responded to more quickly than incongruent trials), which diminishes and fully reverses to contrast effects as prime duration is increased (i.e. congruent trials are responded to more slowly than incongruent trials). Finally, as prime duration is increased even farther, priming effects eventually disappear altogether. The model's qualitative and quantitative fit of the data not only supports the theory of habituation as it applies to affective processing, but it also suggests that perhaps similar neural dynamics are involved in the identification and processing of both words and affective stimuli.

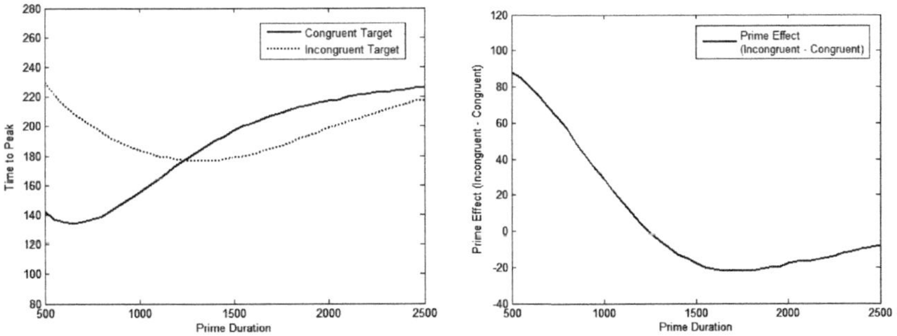

Fig. 5. Simulated data using the proposed affective habituation model. The time to peak response was used as an indicator of relative ease of processing, and directly corresponds to the observed behavioral reaction times. The graph on the left shows the time-to-peak values for congruent trials and incongruent trials separately, while the graph on the right shows the overall "prime effect", or the relative difference between the two priming conditions.

6 General Discussion

The purpose of this chapter was to explore the dynamics underlying affective processing in order to better understand the way in which we integrate and interpret emotional information. As observed by Aristotle over 2,000 years ago, "anybody can become angry - that is easy, but to be angry with the right person and to the right degree and at the right time and for the right purpose, and in the right way - that is not within everybody's power and is not easy." As Aristotle keenly observed, intelligent behavior requires that we respond to the right event, in the right time, and to the right degree. That is, an appropriate emotional response should only consider the relevant events at hand, and not be confused by unrelated recent events. At the same time, events unfold over time and take time to fully process. This creates a dilemma for intelligent behavior in which processes of "persistence" have to be balanced against processes of "separation."

In addition to the problem of separation and proper attribution of events, organisms also face the task of maintaining an ability to respond to a full range of events – sensory events that range, say, in brightness, or emotional events that range in valence . As discussed in prospect theory, adaptation (or habituation) helps maintain this ability (Kahneman & Tversky 1973). Because repeated exposure to a stimulus causes the responsiveness of the sensory system to be diminished over time, habituation brings you back to a baseline level from which you are better at detecting the next change in the environment. For example, adaptation to darkness allows you detect brighter and darker things. Adaptation to happiness, allows you to distinguish events that are better or worse than your current state.

This idea of affective habituation even relates to long term adaptation to significant events over the time course of months or years. The term "hedonic treadmill" was coined by Brickman and Campbell (1971), which captures the tendency of a person to remain at a relatively stable level of happiness despite major life occurrences. For example, Brickman, Coates, and Jonoff-Bulman (1978) reported that there were only small differences in life satisfaction between lottery winners and control subjects, while Silver (1982) found that the affective experience of paraplegics was already predominantly positive after only a few weeks of their accident. These types of studies demonstrate that humans rapidly adapt to their current situation returning to some sort of set affective baseline.

Although we do not claim that the same mechanisms are directly responsible for affective habituation as it occurs over the course of milliseconds and over the course of years, it is interesting to consider some of the apparent parallels that are present. Specifically speaking to those effects of affective habituation that are seen in short term priming, however, we propose a dynamic neural network model. The model takes into account the process of habituation as it affects the perceptual and affective system's ability to parse the stream of incoming information and correctly integrate and interpret the cues at hand. Not only can this model capture the effects of priming that are observed behaviorally, but considering that the proposed automatic dynamics can handle both assimilation and contrast (and the gradual transition between the two), it also suggests that it may be premature to consider more elaborate controlled discounting theories.

It is important to note, however, that while the reported simulation provides a good account of the interaction between priming conditions and prime duration (i.e., a comparison of the right panel of Figure 4 to that of Figure 5), it does not necessarily capture the main effect of prime duration itself (i.e., a comparison of the left panel of Figure 4 to that of Figure 5). As can be observed visually, there is no main effect of prime duration in the behavioral data (i.e., the average of the two conditions is perfectly flat), however, the average of the two conditions in the simulated data exhibits a u-shaped trend. This trend emerges because of inhibition within the layers of processing, which results in a single prime duration that produces the most interference regardless of valence, as compared to other prime durations. The heightened degree of interference at this point restricts the level of activation a target response is able to accumulate, thus resulting in a response that reaches its peak value more quickly. However, at shorter and longer prime durations there are lower levels of inhibition, and therefore, the response to the target is able to reach a higher level, also taking a longer period of time. One possible explanation for this apparent lack of a prime duration effect, is that the model does not currently include an alerting mechanism to know when to respond. And, with the incorporation of such a mechanism, a better fit of the behavioral data could be possible.

Nonetheless, human behavior generally confirms the assumptions made by a habituation account, and the proposed dynamics may be useful in developing an artificial affective system. Other related models have been proposed to account for similar patterns of effects, for example, the psychophysical account by Klauer, Teige-Mocigemba, and Spruyt (2009). This model, however, is adapted from descriptive principles of perceptual processing (e.g., the Weber effect), but it does not specify the underlying mechanism. Therefore, it is limited in its predictive power. For instance, though it makes predictions regarding the manipulation of prime duration, it does not specify the exact dynamics by which the effects occur. Our account, however, proposes that it is specifically lingering activation and neural habituation, which produce various patterns of priming effects.

Whether or not it is specifically the mechanisms currently proposed, any natural or artificial system needs to encompass some sort of specific processing mechanisms, which will allow it to process the continuous stream of incoming information and appropriately adjust its affective responses. Obviously, an "emotional computer" (which will react to users' emotions by adapting its functioning) should have some "knowledge" about how long people stay upset or excited (how fast they adapt), how to structure positive or negative stimuli for maximal impact, and when a possible mis-attribution of emotion is likely to influence a decision or behavior. These traits, attainable with the automatic dynamics of the proposed habituation model, are each relevant for the artificial system to integrate affect with cognition, to learn and communicate in an interactive way, and to behave in an appropriate manner. Thus, by using a formal model that incorporates the same dynamics observed in human behavior, an artificial system's affective processing may become more realistic and appropriate over a wide range of situations.

References

Abbott, L.F., Varela, J.A., Sen, K., Nelson, S.B.: Synaptic depression and cortical gain control. Science 275, 220–224 (1997)

Barlow, H.B., Hill, R.M.: Evidence for a physiological explanation of the waterfall phenomenon and figural after-effects. Nature 200, 1345–1347 (1963)

Brickman, P., Campbell, D.T.: Hedonic relativism and planning the good society. In: Appley, M.H. (ed.) Adaptation-level theory, pp. 287–305. Academic Press, New York (1971)

Brickman, P., Coates, D., Janoff-Bulman, R.: Lottery winners and accident victims: Is happiness relative. Journal of Personality and Social Psychology 36(8), 917–927 (1978)

Chance, F.S., Nelson, S.B., Abbott, L.F.: Synaptic depression and the temporal response characteristics of V1 cells. The Journal of Neuroscience 18, 4785–4799 (1998)

Dijksterhuis, A., Smith, P.K.: Affective habituation: Subliminal exposure to extreme stimuli decreases their extremity. Emotion 2(3), 203–214 (2002)

Fazio, R.H., Sanbonmatsu, D.M., Powell, M.C., Kardes, F.R.: On the automatic activation of attitudes. Journal of Personality and Social Psychology 50, 229–238 (1986)

Glaser, J., Banaji, M.R.: When fair is foul and foul is fair: Reverse priming in automatic evaluation. Journal of Personality and Social Psychology 77(4), 669–687 (1999)

Gotts, S.J., Plaut, D.C.: The impact of synaptic depression following brain damage: A connectionist account of "access/refractory" and "degradedstore" semantic impairments. Cognitive, Affective, & Behavioral Neuroscience 2, 187–213 (2002)

Huber, D.E., Shiffrin, R.M., Lyle, K.B., Ruys, K.I.: Perception and preference in shortterm word priming. Psychological Review 108(1), 149–182 (2001)

Huber, D.E., O'Reilly, R.C.: Persistence and accommodation in short term priming and other perceptual paradigms: Temporal segregation through synaptic depression. Cognitive Science 27(3), 403–430 (2003)

Hurvich, L.M., Jameson, D.: An opponent-process theory of color vision. Psychological Review 64(6), 384–404 (1957); Irwin, K.R., Huber, D.E., Winkielman, P.: As similation and contrast in affective priming: An habituation account (in preparation)

Kahneman, D., Tversky, A.: On the psychology of prediction. Psychological Review 80(4), 237–251 (1973)

Klauer, K.C., Teige-Mocigemba, S., Spruyt, A.: Contrast effects in spontaneous evaluations: A psychophysical account. Journal of Personality and Social Psychology 96(2), 265–287 (2009)

Lombardi, W.J., Higgins, E.T., Bargh, J.A.: The role of consciousness in priming effects on categorization – assimilation versus contrast as a function of awareness of the priming task. Personality and Social Psychology Bulletin 13(3), 411–429 (1987)

Martin, L.L., Seta, J.J., Crelia, R.A.: Assimilation and contrast as a function of peoples willingness and ability to expend effort in forming an impression. Journal of Personality and Social Psychology 59(1), 27–37 (1990)

Meyer, D.E., Schvaneveldt, R.W.: Facilitation in recognizing pairs of words: Evidence of a dependence between retrieval operations. Journal of experimental psychology 90(2), 227–234 (1971)

Murphy, S.T., Zajonc, R.B.: Affect, cognition, and awareness – affective priming with optimal and suboptimal stimulus exposures. Journal of Personality and Social Psychology 64(5), 723–739 (1993)

Nelson, S.B., Varela, J.A., Sen, K., Abbott, L.F.: Functional significance of synaptic depression between cortical neurons. In: Bowers, J. (ed.) Computational neuroscience, pp. 429–434. Plenum Press, New York (1997)

Pantic, L., Torres, J.J., Kappen, H.J.: Associative memory with dynamic synapses. Neural Computation 14, 2903–2923 (2002)

Posner, M.I., Snyder, C.R.R.: Facilitation and inhibition in the processing of signals. In: Rabbit, P.M.A., Dornic, S. (eds.) Attention and performance, vol. V, pp. 669–682. Academic Press, New York (1975)

Rieth, C.A., Huber, D.E.: Using a neural network model with synaptic depression to assess the dynamics of feature-based versus configural processing in face identification. In: Proceedings of the 27th Annual Conference of the Cognitive Science Society, pp. 1856–1861. Erlbaum Associates, Hillsdale (2005)

Silver, R.L.: Coping with an undesirable life event: A study of early reactions to physical disability. Unpublished doctoral dissertation. Northwestern University, Evanston, IL (1982)

Stapel, D.A., Koomen, W., Ruys, K.I.: The effects of diffuse and distinct affect. Journal of Personality and Social Psychology 83(1), 60–74 (2002)

Solomon, R.L., Corbit, J.D.: An opponent-process theory of motivation: Temporal dynamics of affect. Psychological Review 81(2), 119–145 (1974)

Titchener, E.B.: Lectures on the elementary psychology of feeling and attention. McMillan, New York (1908)

Tsodyks, M.V., Markram, H.: The neural code between neocortical pyramidal neurons depends on neurotransmitter release probability. Proceedings of the National Academy of Science 94, 719–723 (1997)

Van de Water, M.: How to see at night. The Science News-Letter 41(23), 358–359 (1942)

Chapter 3
Judgement as to Whether or Not People Are "Involved, Enjoying and Excited," Based on the Visual and Physiological Information

Yoshimasa Ohmoto, Takashi Miyake, and Toyoaki Nishida

Graduate school of Infomatics, Kyoto University
Yoshidahonmachi, Sakyo-ku, Kyoto-shi, Kyoto-hu, Japan
{ohmoto@,miyake@ii.ist.,nishida@}i.kyoto-u.ac.jp

Abstract. We feel uncomfortable when we communicate with Embodied Conversational Agents (ECAs). This is because they are unable to express their mental state on Emotions using their method of communication adequately. Especially in multi-user interaction, ECAs have to detect the mental states of each user in order to interact naturally. In addition, the mental state of a person may be affected by extrinsic factors, such as the states of people around. Physiological indices are useful to understand the mental states of a person. It was impractical for ECAs to measure physiological indices of users. In this study, we experimentally investigated whether we could detect intrinsic "involvement, enjoyment, and/or excitement" ("I-measure") of a person and the extrinsic "I-measure." That was a social atmosphere of "I-measure," using visual information. As a result, both intrinsic and extrinsic "I-measure" could be detected by moving distances and speeds of a user's motion. Each threshold to detect motions was different: The threshold to detect extrinsic "I-measure" was lower than that of intrinsic "I-measure." We then proposed a method to detect the "I-measure" of a person and an atmosphere of "I-measure" in a multi-user interaction.

Keywords: multi-user interaction, mental state detection, physiological indices.

1 Introduction

It is necessary for smooth communication to detect communication conditions and to respond the conditions appropriately. People usually communicate with others with emotional expression. The emotional expression is a strong clue to detect whether communication is smooth or not and to interpret an atmosphere of the communication. It is uncomfortable to disinterestedly communicate with processing personnel and salesmen who have a pasted smile. The reason why we feel uncomfortable in the situation is that the emotional expression does not change along with the communication. We

T. Nishida et al. (Eds.): Modeling Machine Emotions for Realizing Intelligence, SIST 1, pp. 35–52.
springerlink.com

also feel uncomfortable when we communicate with Embodied Conversational Agents (ECAs).

ECAs are expected to be an effective interface for many users, which included computer novices, in diverse fields, such as education, information service, healthcare, or marketing, for example ([3], [14], [16]). There are many which interact with ECAs. Typical issues are natural language communication, operation environment without special devices or sensors, multi-task control, and multi-user interaction. The issue of multi-user interaction has to be solved in order to use ECAs in public spaces. For example, associate friends usually form a group in public spaces. The group then together uses an ECA. The multi-user interactions include many issues which do not occur with one-to-one interactions. Traum [18] listed these issues; which the agent needs to identify the roles of each utterance. These include the speaker, listener, address to be able to appropriately decide when and what to say. In order to solve these issues, we must provide information by using verbal and nonverbal expressions and to recognize conditions of multi-users and the services to be accounted for.

Here, as a first step to understand a social atmosphere, we consider the conditions of multi-users with emotional expressions as one of the tasks involved in multi-user interaction. There are many conditions in this type of interaction. We have focused on the condition where users were in the state of "involved, enjoying or excited." For the remainder of this paper, we will refer to this mental state of the user as "I-measure." One of the reasons why we focused on "I-measure" is that "I-measure" is a strong clue to determine whether the people can interact efficiently or not. Another reason is that states of "I-measure" play an important role in interrupting the interaction with appropriate timing. For example, we do not change the subject of a conversation when the conversation partners are in a state of "I-measure." On the other hand, we may quickly change the subject of a conversation when the social atmosphere of the conversation is not characterized as "I-measure."

In addition, the mental state of a person may be affected by extrinsic factors, such as the states of people around. They often respond differently to the same stimulus depending on the situation. After considering time, place and occasion, it is necessary to generate fitting responses. For example, people usually talk to a silent person during a communication. However, if a person's enjoyment is affected by the social atmosphere that is characterized as "I-measure," the response may be different.

It is also useful for evaluation of an agent to detect "I-measure" state. A questionnaire is a typical and subjective method to evaluate performance. We, however, cannot evaluate the performance in every scene using this method. We may use "I-measure" detection as one of the indices to evaluate the user's response to an agent for every scene because people are interacting efficiently in "I-measure" scene.

To sum up, the purpose of this research is to develop a robust method for detecting intrinsic "I-measure" of a person and extrinsic "I-measure," that was an atmosphere of "I-measure." This may be done by using visual information in multi-user interaction containing emotional expressions. We experimentally investigated using the following three issues; (1) whether a person's "I-measure" could be detected by using visual information, (2) whether the atmosphere of "I-measure" could affect members who were not directly involved in the state of "I-measure," (3) whether the affected "I-measure" could be detected by using visual information. The reliability of subjective evaluation methods was not sufficient because the atmosphere of "I-measure" was ambiguous. We therefore used physiological indices to evaluate the detection of "I-measure" states.

2 Previous Work

Some investigations have focused on emotions and the social atmosphere in interaction.

Gebhard and Kipp [6] reported that the model which they used in the paper provided textual dialog situations with authentic believable emotions and moods. It is difficult to objectively classify all dialog situations. In addition, it is even more difficult to interpret the contexts in multi-party conversations which have complex structures.

Peters et al [15] presented a model of an Embodied Conversational Agent able to establish, maintain and end the conversation based on the ECA's perception of the level of interest. This may be perhaps estimated by using gaze. Eichner et al [4] and Hoekstra et al [7] developed a system which could adapt to the current visual interest state of the user. This may provide a more personalized, context-aware, attentive experience of the presentation by using eye gaze. Multi-party conversations have more complex structures than one-on-one conversation or presentation. Therefore, the agent may change its expression of eye gaze according to change in the conversation structures including multi-party conversations.

Wrede et al [19] reported that "involvement" could be characterized by deviations in fundamental frequency and energy. It was pointed out that prosodic verifiable features were significant cues to detect "I-measure." However, prosodic features dynamically change the relationship between a speaker and a listener. In addition, sound source separation is difficult during multi-party conversation. We did not focus on prosodic features because we assumed that the visual information would also change in the situation where prosodic features change.

Burkitt and Romano [2] presented a computational model for believable adaptable, social intelligent characters created based on established psychological theories, relating mood, memory, personality and emotion. Kang et al [9] explored associations between the five-factor personality traits of human subjects and their feelings of rapport when they interacted with a virtual agent or humans. In these investigations, they did not discuss any method to detect moods, emotions, personality, for example.

Morency et al [13] discussed how sequential probabilistic models could automatically learn from a database of human-to-human interactions to predict listener backchannels. This was done by using the speaker multi-modal output features of prosody, spoken words and eye gaze. However, it was not revealed how the backchannels affect the moods and emotions.

In the work described above, "I-measure" was evaluated by using subjective reports of participants and experimenters. We need a method to objectively detect "I-measure." Ishii and Nakano [8] proposed a conversational engagement estimation algorithm that estimates the user's degree of engagement by use of gaze transition patterns. When the engagement is high, the participants may be in the state of "I-measure." It is an indirect method to detect "I-measure." Physiological indices are used to directly detect "I-measure."

The relationships between physiological indices and mental states are well-known. In addition, some researches reported that it was useful to understand the mental states in interaction. For example, Mandryk et al [12] reported that the normalized physiological indices would correspond to reported subjective experience. We believed that "excitement," "stress," and "concentration" are related to the state of "I-measure" in this research. We therefore evaluated the "I-measure" detection using physiological indices which were related to the mental states.

Some agents used physiological indices to understand the user's mental states. Becker et al [1] proposed an agent which analyzed physiological user information to trigger empathic reactions of the ECA during game play, there by increasing the level of social engagement. Prendinger and Ishizuka [17] developed an interface application which measured physiological data such as skin conductance and electromyography of user in real-time and interpreted them as emotions. They addressed the user's states in the form of empathic feedback. Moreover, they presented results from an exploratory study which aimed to evaluate the impact of the Empathic Companion by measurement of users' skin conductance and heart rate. The outcome of the experiment suggests that empathic feedback has a positive effect on the subject's stress level while listening to the interviewer's questions.

We have discussed the achievements and limitations of previous works. Some researchers presented an agent able to adapt to the current state of the user. Few contributions, however, have investigated an agent's behavior and a way to detect the state of users in a multi-user interaction. In some investigations, the state of the user was detected by the speech context, prosodic features, or by a direct use of physiological indices. It is, however, difficult to implement these elements in an agent, which is used in real-world interaction. This is because they induce the conditions to the measuring environment and the users' interactions.

This study differs from previous works in following points. We have focused on "I-measure." That was one of the important states of the user in a multi-user interaction. We also experimentally investigated the intrinsic "I-measure" of an individual and also extrinsic "I-measure" related to the social atmosphere. We also used physiological indices to obtain reliable evaluations.

The remainder of this chapter is organized as follows. Section 3 describes "I-measure" as used in this research and the methods used to detect it. Section 4 explains the settings of the conducted experiments. Section 5 describes our analyses of the way to detect "I-measure" and effects of a social atmosphere related to "I-measure." We then discuss future work and the method of implementations for an ECA to detect states of "I-measure." Section 6 provides the discussions and conclusions.

3 Approach

We investigated whether "I-measure" could be detected by using visual information. For the investigation, we evaluated the detection based on visual information obtained by using physiological indices. In this section, we describe how to detect the "I-measure" by visual information and physiological indices.

For this study, we have defined "I-measure" as "mental states when people are activated by a mood in the interaction, such as Involvement, enjoyment, excitement (I-measure)."

3.1 The Method to Detect "I-Measure" by Using Visual Information

Involvement, enjoyment and excitement lead to vigorous body movements of a person. On the other hand, people are slow to respond to stimuli or keep still when they are not involved, enjoying the situation, or excited. We believe that people can detect "I-measure" of their communication partners on the basis of vigorous movements.

We describe below the method by which one can detect "I-measure" of an individual and a social atmosphere of "I-measure."

The method to detect "I-measure" of an individual. The "I-measure" state of an individual is detected by observing the speed and distance of face, head, hand and shoulder movements. The speed and distance of these movements differ between individuals. Detectors therefore have to observe how fast and how long persons who are exhibiting an "I-measure" state move their bodies when they are deemed to exhibit "I-measure" in a communication. For this, the observers will watch a video of the subjects several times. After this observation they may, they detect the "I-measure" states in the person.

Voice information, such as pitch and power, also provides clues to detect "I-measure." It is, however, difficult to separate sound sources when several users talk nearby with each other. We expected that we could detect

"I-measure" by using visual information because factors of multi-user interactions. These may be turn taking during a conversation, or change the topic, trigger body movements of the participants in the communication.

We considered using facial expressions to detect "I-measure." We were able to detect basic emotions, such as pleasure, sadness for example. Ekman [5] said that we had to pay attention to subtle changes in facial expressions in order to detect complex and ambiguous mental states. We could only consider small numbers of those subtle changes in the preliminary experiment. Moreover, those subtle changes varied with the individual. We did not focus on facial expressions in an attempt to detect "I-measure."

A method which can be used to detect the atmosphere "I-measure". A social atmosphere related to "I-measure" is determined if more than half of the participants in a communication are in a state of "I-measure." The level of the atmosphere is calculated as an average of the levels of the participants' "I-measure" states. For example, when three out of four participants are in a state of "I-measure," the level of the social atmosphere of "I-measure" is determined by taking the average of the level of the three components of "I-measure" states.

3.2 The Method Used to Detect "I-Measure" by the Use of Physiological Indices

Each component which includes the following involvement, enjoyment and excitement can be detected by using some physiological indices. We now briefly explain of the advantages and disadvantages found when using the physiological indices. After that, we describe the physiological indices which we used in this research, and the method employed to detect states of "I-measure," which are based on those indices.

Some advantages and disadvantages of physiological indices. Physiological indices include biological reactions, such as brain waves, potential differences in cardiography, variations in blood pressure, pulse waves, respiration, body temperature, muscle potential, skin conductance for example. These reactions reflect the mental state of a person. We can use them as indices representing the person's mental state.

It is noted that physiological indices have advantages as listed below.

- **We can measure these states as physical quantities**
  Psychological activities can be quantified as physical quantities, which include variation of voltage, frequency as examples. In addition, we can easily analyze the data to a certain degree.
- **We can record physiological these reactions in real time**
  By this means it is possible to analyze the variability of physiological indices over a period of time as it is recorded as time-series of data in real

time. Questionnaires are usually done to indicate personal psychological activities following some activities.
- **We can observe mental states of a person**
  In some cases, physiological indices could detect mental states without a change in facial expressions on actions.

Physiological indices have certain disadvantages as indicated below.

- **Physiological indices themselves may contain noise caused by external stimuli, and may be dependent on light conditions, physical conditions, age of the subject and other factors**
  External factors, such as the light condition, sound, individual differences. These factors can affect the physiological responses. They may not affect the outcome.
- **Data of physiological indices may contain noise and artifacts caused by alternating electromagnetic fields interference**
  Alternating electromagnetic fields, noise or other artifacts can be caused by a movement of the lead wires used to measure of physiological indices.
- **The presence of sensors or the restriction of movement may possibly inhibit on distant natural communication**
  The appearance of a person changes if sensors are put on the body to record physiological indices. In addition, it is better not to change the position of the sensors to reduce the noise. These factors may inhibit natural communication.

The physiological indices used in this study. Skin Conductance Response and Respiration were used as Physiological Indices which were considered as reflecting a person's mental state.

"Skin conductance response (SCR)" which is the skin conductance caused by emotional sweating. This can occur due to excitement and mental stress and concentration as increase. Respiration becomes more rapid and is caused by a mental state. We can detect laughing by rapid respiration. These features show how we can detect states of "I-measure" by using skin conductance response and respiration.

We had considered whether to use brain waves or pulse waves. Brain waves were not used as the measurement of brain waves prevented participants from using natural communication. This was because we had to fixate the head piece to the head the participant to measure the brain waves. The reason for not using the pulse wave was that the data of the pulse wave contained a large amount of noise.

The method used to detect "I-measure" in this study. We defined the criteria of "Increase of SCR" and "Rapid Respiration" as follows:

- The Increase in SCR: The SCR increases of more than 5
- Rapid Respiration: Peaks of respiration are observed more than four times every three seconds.

The criterion "Increase in SCR" was defined and based on previous work given in reference [11]. The criterion "Rapid Respiration" was defined so as to also detect the change of respiration other than that obtained by laughing, and to avoid the detection of changes associated with normal communication.

4 Experimental Methods Used to Record Responses in "I-Measure"

We conducted an experiment to record the responses of participants in "I-measure" situation on videos and using physiological indices. We describe the setting, the task and the procedures employed in the experiment.

4.1 Participants

Four participants composed a group and this will be referred as a "quartet" in what follows. Each quartet included undergraduate students who were 21 or 22 years old and were acquainted with each other. A total of three quartet of male which consisted students participated.

4.2 The Task

The participants were asked to answer a series of quiz questions. Free communication between the students was permitted. A quiz session were consisted of ten questions. The participants attended three sessions. That is they answered a total of 30 questions.

4.3 Experimental Setting

An Embodied Conversational Agent introduced a series of quiz questions. This is because we investigate a method to detect "I-measure" in multi-user interaction in which an ECA was involved. The agent is referred to as the Quiz Agent in the following. A 100-inch screen displayed the Quiz Agent. The experimenter could make the quiz agent start a quiz or accept an answer to the quiz by using a notebook computer. The experimenter sat out of participants' view.

Participants were sitting on chairs to avoid introducing disturbances into the data. They were seated in the form of an arch in front of the screen (Fig. 1). They were able to watch the screen and communicate to one and another naturally.

A video camera which was placed on top of the screen recorded the participants' behavior. A "Polymate" is a device used to measure the physiological indices of the participants'. The Polymate was placed on a table out of sight behind the participants.

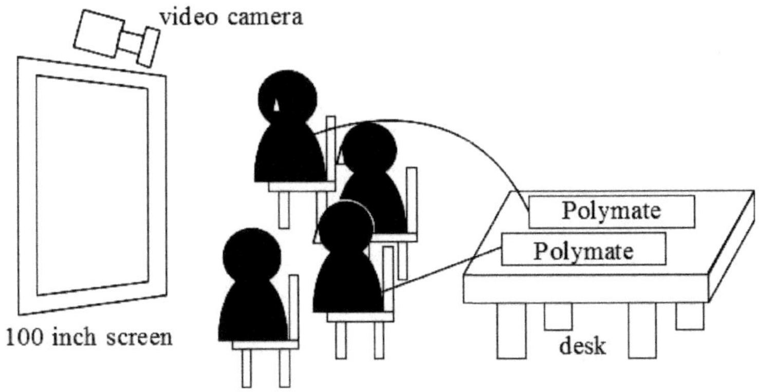

Fig. 1. Experimental Arrangement

We recorded two participants who were seated on the chairs. In this experimental setting, the participants who sat next to each other tended to talk together. The measuring participants were separated to avoid that they communicated in only one conversational field.

Electrodes to measure SCR were placed on a forefinger and a ring finger of a left hand. The left arm was place on an armrest to support the left hand. Respiration was measured by a tube placed around the chest. Participants did not experience discomfort from this tube.

4.4 Procedure

The experiment was conducted as follows:

1. The experimenter provided instructions on the experiment. He did not explain the ultimate objective.
2. The experimenter allocated where participants must sit.
3. Two participants were fitted with electrodes to measure SCR and the tube to measure respiration. They asked to place their arm on an armrest. There were no more restrictions.
4. The experimenter began record using the video camera and the Polymates. He started the Quiz Agent to initiate the first quiz session.
5. The participants answered all the quiz questions in order discussed. The order of the questions was determined in advance.
6. After the participants finished a quiz session, the experimenter stopped the Quiz Agent, the video camera and the Polymates.
7. The participants then had a three-minute break. They were remained connected to the electrodes and the respiration tube. They could talk to each other.
8. The steps 4-7 were repeated three times. The questions in each session were different. The participants answered 30 questions in total.

Participants answered the quiz questions for about 45 minutes. The duration of three session for Group A was 50 minutes, that for Group B was 40 minutes, and that for Group C was 50 minutes.

5 Analyses of Effects of an Atmosphere of "I-Measure"

We analyzed the data, which consisted of visual information and physiological indices, in following manner.

Firstly, persons segmented video data independently, and then annotated whether a target person and the social atmosphere were a state of "I-measure" based on the visual information. We call the persons as "coders." Secondly, the coders performed this procedure on the physiological indices data. For these parts of the analysis, the annotation tool, iCorpusStudio [10] was used. The three analyses were done to investigate the possibility of detecting states of "I-measure" by using visual information, and to identify effects of a social atmosphere in "I-measure." Fig. 2 shows an example of the annotated data. In this example, when participant B was in "I-measure," the atmosphere was also in "I-measure." We observed increasing SCR of B and the shallow respiration of B. We explain the details of each step in the procedure and show the results of the analyses below.

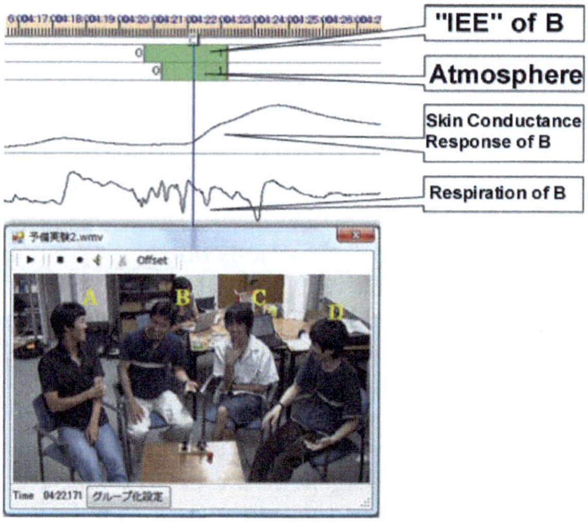

Fig. 2. Showing One Annotation Data of the Groups

5.1 The Annotations of "I-Measure" by Using Visual Information

We have already explained the method used to detect "I-measure" for a person by using videos in Section 3.1. Coders annotated the level of the "I-measure" state on a scale of Zero to One or of One to Five. They segmented the videos into "I-measure" units, where the level of the "I-measure" state was the same. Consecutive "I-measure" units did not have to be annotated the consecutive level of the "I-measure." For example, coders could annotate 1/4/1/3 in four consecutive "I-measure" units.

A level of a social atmosphere of "I-measure" was determined and based on whether more than two out of four participants are considered to be in a state of "I-measure." The level of the social atmosphere related to "I-measure" was determined by the average of the levels of the participants' "I-measure" states.

Two coders, the experimenter and a person who often made video annotation (we refer to the coder as a "reference coder" below), segmented the videos and annotated the level of the "I-measure" state.

We compared the positions of annotations of the two coders to confirm their reliability. Table 1 shows the result. In 82% of the 214 annotations, the positions of the two coders' annotations overlapped. More than half were between the two coders' annotations. This enabled us to confirm the reliability of the positions of the annotations.

Table 1. Confirmation of the Positions of Annotation Reliability

Number of Annotations	Matched Positions	Mismatched Positions
214	176 (82%)	38 (18%)

In addition, we compared the levels of the annotations on a scale of one to five to further confirm the reliability. Table 2 shows this result. 70% annotations were completely matched the level between two coders. The chance level was 20%. Therefore, we could confirm the reliability of the levels of the annotations to some extent.

Table 2. Confirmation of the Level of Annotation Reliability

Number of Annotations	Complete Match	Difference 1	Difference 2	Difference 3
176	123	44	9	0
Percentage	70%	25%	5%	0%

5.2 The Annotations of "I-Measure" Obtained by the Use of Physiological Indices

We automatically determined whether a participant was in a state of "I-measure" at each of the "I-measure" units. These were segmented by the use of a coder. This determined the "increase of SCR" and the "rapid respiration." We did not consider the levels of the "I-measure" state in these determinations.

Table 3. The Percentage of Laughing and Other Behavior which Exhibited "Rapid Respiration"

"Rapid Respiration"	Laughing	Other
113	79 (70%)	34 (30%)

Usually when people laugh, their respiration is rapid. We therefore determined that most of persons who laughed were in a state of "I-measure." We confirmed the percentage of "laughing" in all annotations. Here the "rapid respiration" criterion was used to determine that the participants were in a state of "I-measure." Table 3 shows the result. The participants were laughing in 70% of 113 annotations. Therefore, we could confirm that "rapid respirations" did not completely correspond to "laughing." There was however a high correlation with "laughing."

5.3 Analysis of Accuracy of "I-Measure" Detection by Using Visual Information

We believe it is possible detect "I-measure" states in people by using visual or phonetic information. It is, however, not obvious. We investigated whether people could detect the state of "I-measure" based on visual information. For this investigation, we compared the annotations which had an "I-measure" level on a scale from Zero to One, based on visual observation with that obtained by physiological indices.

Table 4 shows the result. The average concordance rates between annotations by visual information and those by physiological indices. They were 74% in Group A, 60% in Group B, 67% in Group C. It was therefore concluded that some people could detect better than others the "I-measure" states by using visual information to some extent.

Conversely, the concordance rates were not so high, especially not for the non-"I-measure" annotations. In most cases, annotations by visual information indicated non-"I-measure" states. Annotations based on physiological indices indicated "I-measure" states. We conclude that low-level "I-measure" states were difficult to detect by using visual information. That could be detected by using physiological indices.

5.4 The Analysis of Sensitivity of "I-Measure" Detection When Using Visual Information

We investigated whether low-level "I-measure" states, which were difficult to detect on the basis of visual information, could be detected using physiological indices. For this investigation, we classified annotations, which had a level of "I-measure" on a scale from One to Five by using visual information, divided into the two groups of "I-measure" state or non-"I-measure" state

Table 4. The Concordance Rates between Annotations using Visual Information and Ones using Physiological Indices

(a) Group A

	Number of "I-measure" using Physiological Indices	Number of non-"I-measure" using Physiological Indices
Number of "I-measure" using Visual Information	96 (83%)	20 (17%)
Number of non-"I-measure" using Visual Information	42 (35%)	78 (65%)

(b) Group B

	Number of "I-measure" using Physiological Indices	Number of non-"I-measure" using Physiological Indices
Number of "I-measure" using Visual Information	38 (76%)	12 (24%)
Number of non-"I-measure" using Visual Information	30 (57%)	23 (43%)

(c) Group C

	Number of "I-measure" using Physiological Indices	Number of non-"I-measure" using Physiological Indices
Number of "I-measure" using Visual Information	44 (86%)	7 (14%)
Number of non-"I-measure" using Visual Information	29 (52%)	27 (48%)

according to a threshold. The threshold ranged from Two to Five. For example, when the threshold was three, annotations that were rated one or two were classified as "I-measure" state and annotations that were rated three, four or five were classified as non-"I-measure" state. After the classification process, we annotated the level of "I-measure" again on a scale from Zero to One according to the groups. We call this annotation as "re-annotation." We then compared the re-annotated data with annotations obtained by using physiological indices.

Table 5 lists the results. The average concordance rates in Group A were 50% with threshold of five, 60% with threshold of four, 63% with threshold of three, and 73% with threshold of two. Those in Group B were 40% with threshold of five, 45% with threshold of four, 60% with threshold of three, and 76% with threshold of two. Those in Group C were 55% with threshold of five, 51% with threshold of four, 67% with threshold of three, and 77% with threshold of two. In all of these cases, the average concordance rates are highest for the threshold of two. The concordance rates of non-"I-measure" states are higher than the values in table 4. Therefore, we could confirm that low-level "I-measure" states which were difficult to detect by using visual information could be detected by using physiological indices. We suggest that the low-level "I-measure" states may be detected by using visual information with a low threshold.

Table 5. The Average Concordance Rates for Particular Levels of Annotations

(a) Group A

Threshold		Number of "I-measure" using Physiological Indices	Number of non-"I-measure" using Physiological Indices
5	Number of "I-measure" using Visual Information	6 (100%)	0 (0%)
	Number of non-"I-measure" using Visual Information	11 (100%)	0 (0%)
4	Number of "I-measure" using Visual Information	22 (96%)	1 (4%)
	Number of non-"I-measure" using Visual Information	22 (76%)	7 (24%)
3	Number of "I-measure" using Visual Information	62 (90%)	7 (10%)
	Number of non-"I-measure" using Visual Information	47 (64%)	27 (36%)
2	Number of "I-measure" using Visual Information	109 (75%)	37 (25%)
	Number of non-"I-measure" using Visual Information	45 (30%)	105 (70%)

(b) Group B

Threshold		Number of "I-measure" using Physiological Indices	Number of non-"I-measure" using Physiological Indices
5	Number of "I-measure" using Visual Information	4 (80%)	1 (20%)
	Number of non-"I-measure" using Visual Information	8 (100%)	0 (0%)
4	Number of "I-measure" using Visual Information	18 (86%)	3 (14%)
	Number of non-"I-measure" using Visual Information	22 (96%)	1 (4%)
3	Number of "I-measure" using Visual Information	38 (76%)	12 (24%)
	Number of non-"I-measure" using Visual Information	30 (57%)	23 (43%)
2	Number of "I-measure" using Visual Information	75 (69%)	34 (31%)
	Number of non-"I-measure" using Visual Information	22 (18%)	103 (82%)

(c) Group C

Threshold		Number of "I-measure" using Physiological Indices	Number of non-"I-measure" using Physiological Indices
5	Number of "I-measure" using Visual Information	6 (100%)	0 (0%)
	Number of non-"I-measure" using Visual Information	10 (91%)	1 (9%)
4	Number of "I-measure" using Visual Information	17 (89%)	2 (11%)
	Number of non-"I-measure" using Visual Information	21 (87%)	3 (13%)
3	Number of "I-measure" using Visual Information	44 (86%)	7 (14%)
	Number of non-"I-measure" using Visual Information	29 (52%)	48 (27%)
2	Number of "I-measure" using Visual Information	87 (72%)	33 (28%)
	Number of non-"I-measure" using Visual Information	22 (18%)	103 (82%)

5.5 Analyses of Effects of an Atmosphere of "I-Measure"

When the multi-user interaction has a social atmosphere of "I-measure," it intuitively affects members who have not adopted the "I-measure" state. We then investigated whether the atmosphere of "I-measure" could affect those members who are not directly involved in the "I-measure" state and whether participants could detect the "I-measure" state by using visual information. For the investigation, we made "re-annotation" as described in Section 5.4. After that, we determined the level of the social atmosphere of "I-measure" as defined in Section 3.1. Finally by using visual information for each threshold, we selected scenes in which a participant was not in a state of "I-measure" but in a social atmosphere of "I-measure." We then compared the "re-annotation" in the scenes with annotations based on physiological indices. We could confirm that participants were affected by a social atmosphere of "I-measure," which was difficult to detect based on visual information when annotations based on physiological indices indicated that a participant was in a state of "I-measure" in the selected scenes. Moreover, our findings suggested that people could detect the affected "I-measure" state by using visual information when annotations of "I-measure" by physiological indices in the scenes were decreased after changing the threshold.

Table 6 shows the results. Annotations of "I-measure" by physiological indices in the scenes decreasing as the threshold was decreasing; the threshold

Table 6. The Effects of the Atmosphere of "I-measure"

(a) Group A

Threshold	Number of "I-measure" using Physiological Indices	Number of non-"I-measure" using Physiological Indices
5	4 (100%)	0 (0%)
4	10 (73%)	2 (17%)
3	24 (73%)	9 (27%)
2	18 (38%)	29 (62%)

(b) Group B

Threshold	Number of "I-measure" using Physiological Indices	Number of non-"I-measure" using Physiological Indices
5	2 (50%)	2 (50%)
4	9 (90%)	1 (10%)
3	11 (48%)	12 (52%)
2	4 (13%)	27 (87%)

(c) Group C

Threshold	Number of "I-measure" using Physiological Indices	Number of non-"I-measure" using Physiological Indices
5	1 (50%)	1 (50%)
4	2 (33%)	4 (67%)
3	13 (43%)	17 (57%)
2	9 (26%)	24 (76%)

of two minimized the difference between determinations based on visual information and physiological indices. Therefore, we confirmed that the social atmosphere of "I-measure" could affect participants who were not directly involved in a state of "I-measure." In addition, we could suggest that the affected "I-measure" state could be detected by using visual information with a low threshold.

5.6 Future Work

In future work, we will develop a method of detecting other states of a user in multi-user interactions based on visual information, and evaluate it by using physiological indices. In addition, we will implement the method to an ECA, and experimentally evaluate the effectiveness of the method.

One possible implementation of the method to detect "I-measure" is the following: First, a region of a user in the camera images is detected by image processing, such as background subtraction. Second, moving distances and speed of the user's body motions in the region are detected by an image processing method, such as optical flow. Third, it is judged that the user is in a state of "I-measure" when the moving distances and speed are above a threshold. Fourth, the level of the social atmosphere is determined when it is judged that one or more members of the users group are not in an "I-measure" state. Fifth, the non-"I-measure" members are judged again with a low threshold to detect an affected "I-measure" state when the level of the social atmosphere is above a threshold. We predict that this method would be able to detect about 70% of the "I-measure" states.

For robust and accurate detection of "I-measure" states it is necessary to pay attention to voice information and facial expressions. In order to use them, they have to be measures in a multi-user interaction. However, no system can measure them in a natural interaction. This means that interaction partners do not need to wear any special devices. This is a future work for researchers who investigate a measurement system.

6 Discussions and Conclusions

The intention of this research is the development of a method to detect a level of "I-measure" states in multi-user interaction. We investigated three issues; (1) whether a person's "I-measure" could be detected by using visual information, (2) whether the social atmosphere of "I-measure" could affect members who were not directly involved in a state of "I-measure," (3) whether the affected "I-measure" state could be detected by using visual information. Based on our analyses, we make the following suggestions.

(1) The method by using visual information and the method by using physiological indices resulted in different sensitivities for detecting states of "I-measure." It was more difficult to detect by using visual information than

by using physiological indices. It is noted that the sensitivities required to detect “I-measure” states may differ those obtained from subjective feelings.

(2) The social atmosphere of “I-measure” could affect members not directly involved in the state of “I-measure.” We had to use highly sensitive methods, such as physiological indices, to detect the affected “I-measure” state where the sensitivity was low. However, it is not sufficient to detect these states by using physiological indices. This is because physiological indices do not provide context information.

(3) The affected “I-measure” state could be detected by using visual information having a low threshold. In this study, detection using the lowest threshold was found to be the best. However, the best threshold varies from states to states.

We suggest that one method could be used to detect the “I-measure” state of a person and the social atmosphere of “I-measure” would be based on visual information in a multi-user interaction. This is only the first step necessary to develop ECAs capable of naturally communicating with humans and able to detect the social atmosphere of the communication by detecting emotional condition.

References

1. Becker, C., Prendinger, H., Ishizuka, M., Wachsmuth, I.: Evaluating Affective Feedback of the 3D Agent Max in a Competitive Cards Game. In: Tao, J., Tan, T., Picard, R.W. (eds.) ACII 2005. LNCS, vol. 3784, pp. 466–473. Springer, Heidelberg (2005)
2. Burkitt, M., Romano, D.M.: The Mood and Memory of Believable Adaptable Socially Intelligent Characters. In: Prendinger, H., Lester, J.C., Ishizuka, M. (eds.) IVA 2008. LNCS (LNAI), vol. 5208, pp. 372–379. Springer, Heidelberg (2008)
3. Cassell, J., Sullivan, J., Prevost, S., Churchill, E. (eds.): Embodied Conversational Agents. The MIT Press, Cambridge (2000)
4. Eichner, T., Prendinger, H., André, E., Ishizuka, M.: Attentive Presentation Agents. In: Pelachaud, C., Martin, J.-C., André, E., Chollet, G., Karpouzis, K., Pelé, D. (eds.) IVA 2007. LNCS (LNAI), vol. 4722, pp. 283–295. Springer, Heidelberg (2007)
5. Ekman, P.: Telling Lies: Clues to Deceit in the Marketplace, Politices, and Marriage. W.W. Norton & Company (1985)
6. Gebhard, P., Kipp, K.H.: Are Computer-Generated Emotions and Moods Plausible to Humans? In: Gratch, J., Young, M., Aylett, R.S., Ballin, D., Olivier, P. (eds.) IVA 2006. LNCS (LNAI), vol. 4133, pp. 343–356. Springer, Heidelberg (2006)
7. Hoekstra, A., Prendinger, H., Bee, N., Heylen, D., Ishizuka, M.: Highly Realistic 3D Presentation Agents with Visual Attention Capability. In: Butz, A., Fisher, B., Krüger, A., Olivier, P., Owada, S. (eds.) SG 2007. LNCS, vol. 4569, pp. 73–84. Springer, Heidelberg (2007)

8. Ishii, R., Nakano, Y.I.: Estimating User's Conversational Engagement Based on Gaze Behaviors. In: Prendinger, H., Lester, J.C., Ishizuka, M. (eds.) IVA 2008. LNCS (LNAI), vol. 5208, pp. 200–207. Springer, Heidelberg (2008)
9. Kang, S.-H., Gratch, J., Wang, N., Watt, J.H.: Agreeable People Like Agreeable Virtual Humans. In: Prendinger, H., Lester, J.C., Ishizuka, M. (eds.) IVA 2008. LNCS (LNAI), vol. 5208, pp. 253–261. Springer, Heidelberg (2008)
10. Kijima, H., Bono, M., Sumi, Y., Nishida, T.: Development of the Environment for Multimodal Interaction Analysis, IPSJ SIG technical reports, 2007(99), pp. 63–70 (2002)(Japanese)
11. Lin, T., Omata, M., Hu, W., Imamiya, A.: Do physiological data relate to traditional usability indexes? In: Proceedings of the 17th Australia conference on Computer-Human Interaction, pp. 1–10 (2005)
12. Mandryk, R.L., Inkpen, K.M.: Physiological Indicators for the Evaluation of Co-located Collaborative Play. In: Proceedings of the 2004 ACM conference on Computer supported cooperative work, pp. 102–111 (2004)
13. Morency, L.-P., de Kok, I., Gratch, J.: Predicting Listener Backchannels: A Probabilistic Multimodal Approach. In: Prendinger, H., Lester, J.C., Ishizuka, M. (eds.) IVA 2008. LNCS (LNAI), vol. 5208, pp. 176–190. Springer, Heidelberg (2008)
14. Nishida, T. (ed.): Conversational Informatics. An Engineering Approach. John Wiley and Sons, Chichester (2007)
15. Peters, C., Pelachaud, C., Bevacqua, E., Mancini, M., Poggi, I.: A Model of Attention and Interest Using Gaze Behavior. In: Panayiotopoulos, T., Gratch, J., Aylett, R.S., Ballin, D., Olivier, P., Rist, T. (eds.) IVA 2005. LNCS (LNAI), vol. 3661, pp. 229–240. Springer, Heidelberg (2005)
16. Prendinger, H., Ishizuka, M. (eds.): Life-Like Characters. Tools, Affective Functions, and Applications, Cognitive Technologies. Springer, Heidelberg (2004)
17. Prendinger, H., Ishizuka, M.: The empathic companion: a character-based interface that addresses users' affective states. Applied Artificial Intelligence 19(3/4), 267–285 (2005)
18. Traum, D., Marsella, S., Gratch, J., Lee, J., HartHolt, A.: Multi-party, Multi-issue, Multi-strategy Negotiation for Multi-Modal Virtual Agents. In: Prendinger, H., Lester, J.C., Ishizuka, M. (eds.) IVA 2008. LNCS (LNAI), vol. 5208, pp. 117–130. Springer, Heidelberg (2008)
19. Wrede, B., Shriberg, E.: Spotting "Hot Spots" in Meetings: Human Judgments and Prosodic Cues. In: EUROSPEECH 2003, pp. 2805–2808 (2003)

Chapter 4
Modelling Interaction Dynamics during Face-to-Face Interactions

Yasser Mohammad and Toyoaki Nishida

Graduate School of Informatics,
Kyoto University

Abstract. During face to face interactions, the emotional state of each participant is greatly affected by the behavior of other participants and how much this behavior conforms with common protocols of interaction in the society. Research in human to human interaction in face to face situations has uncovered many forms of synchrony in the behavior of the interacting partners. This includes factors as body alignment, entrainment of verbal behavior. Maintenance of these kinds of synchrony is essential to keep the interaction natural and to regulate the affective state of the interacting partners.

In this chapter we examine the interplay between one partner's use of interaction protocols, maintenance of synchrony and the emotional response of the other partner in the two way interactions.

We will first define the notion of interaction protocol and relate it with the Reactive Theory of Intention and Low Level Emotions. We will then show empirically that the use of suitable interaction protocols is essential to maintain a positive emotional response of the interaction partner during face to face explanation situations. The analysis in this section is based on the H^3R [1] interaction corpus containing sixty six human-human and human-robot interaction sessions. This interaction corpus utilizes physiological, behavioral and subjective data.

Using this result, it is necessary to model not only the affective state of the interacting partners but also the interaction protocol that each of them is using. Human-Robot interaction experiments can be of value in analyzing the interaction protocols used by the partners and modelling their emotional response to these protocols.

We used Human-Robot interactions in explanation and collaborative navigation tasks as a test-bed for our analysis of interaction protocol emergence and adaptation.

The first experiment analyzes how the requirement to maintain the interaction protocol and synchrony restricts the design of the robot and how did we meet these restriction in a semi-autonomous miniature robot. We focus on how low level emotions can be used to act as a mediator between Perception and Behavior.

The second experiment explores a computational model of the interaction protocol and evaluates it in an explanation face to face scenario.

T. Nishida et al. (Eds.): Modeling Machine Emotions for Realizing Intelligence, SIST 1, pp. 53–87.
springerlink.com

The chapter also provides a critical analysis of the interplay between interaction protocols and the emotional state of interaction partners.

Keywords: Interaction Dynamics, Embodied Interactive Control Architecture.

1 Introduction

Producing intelligent agents that have cognitive characteristics similar to that of humans is one of the major goals of research in embodied agents and cognitive robotics. Intelligence is usually defined by AI researchers in behavioral terms. For example Russel and Norvig define an intelligent agent as "*a system that perceives its environment and takes actions which maximize its chances of success*" [2]. This means that the focus of intelligence is behavior or at least the way in which to measure intelligence is to observe system behavior. The mechanism to generate this *intelligent* behavior is not completely irrelevant as argued by Searle in his Chinese room argument [3]. Here, we use the term cognition to represent the totality of mental functions that lead to intelligent behavior. According to this definition, affect and emotion are treated as parts of cognition rather than the opposite of it. Daniel Goleman and other researchers have argued that emotion is just as important in realizing intelligence as rational problem solving. We will show that during face to face interactions it is necessary to model a third component which is the Interaction Protocol.

To understand the interplay between these three cognitive elements we need to consider the final measure of intelligence which is the behavior. How does cognition generate behavior?

Behavior generation in robotics and embodied agents usually takes one of two routes. Either "top-down" or "bottom-up". Traditional BDI agents use a top-down approach. This is where a master plan at one of the layers activates sub-plans in an immediate lower layer and this proceeds until the Action Generation Layer is reached. Most hybird reactive-deliberative architecture systems use a similar technique by having the deliberative layer dictate which of the reactive reactive processes to run [4]. An exception to this procedure is the system proposed in [5]. Here the reactive layer works as an advisor for a deliberation layer making behavior generation go from bottom upwards.

Behavior generation direction must not be confused by the direction of information flow in the architecture. Information flow always contains bottom-up and top-down directions. The main difference between the architectures in this matter is the length of the information loop. In reactive systems, such as the Subsumption architecture, information passes from sensors to actuators through short term passes involving various processes in the robot. In classical GOFAI systems information paths are longer going from the sensors all the way up to the planning layer and then going down to the actuators. This is one of the reasons reactive systems are more successful in Time-Critical Control. Human Behavior Generation in the other hand appears to have both top-down and bottom-up directions. It also combines reactive interaction with deliberative control as shown in Fig. 3.

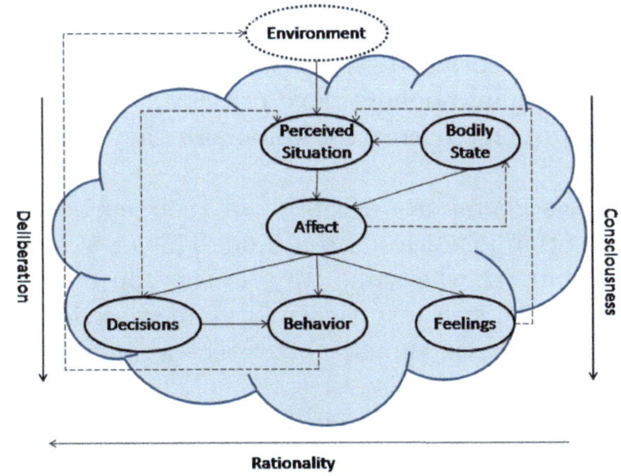

Fig. 1. The full agent cognitive model we utilize. Fig. 3 represents the same information after omitting the rational (left) side of this Figure.

Fig. 1 shows the Behavior Generation Causal Model proposed. This model contains both bottom-up (e.g. a bodily state causing an affect state which in turn activates a specific feeling) and top-down (e.g. a feeling that affects the perceived situation) behavior generation models. It also contains both reactive processes such as "Perceived Situation – Affect – Bodily State" loop and deliberative processes such as "Perceived Situation, Cognition, Decision" loop.

This chapter begins by introducing the relationship between emotion and interaction in section 2. Section 3 defines two approaches used to model cognitive components. Section 4 describes the technique of modeling the low level affective state. It also provides experimental support of the efficacy of the method. Section 2.4 provides details concerning modeling interaction protocols using the two approaches presented in section 3. It also compares the results obtained. This concludes the chapter.

2 Emotion and Interaction

2.1 Theories of Emotion

What is emotion? This question has not easy answer. It is easy to say that fear is an emotion. It is not easy to define the set of common properties that combine fear, happiness, sadness, anger and other feelings and states that we call emotions. Paul Griffiths [6] argued that it is unlikely that all the psychological states and processes that exist in the vernacular category of emotion are sufficiently similar to one another to allow a unified scientific psychology of emotions. He based his claim on noticing that the psychological, neuroscientific and biological theories that best explain any particular subset of human emotions are unable to adequately explain all human emotions. Furthermore, some researchers have argued that even a single emotion such as love requires different theories to explain it depending in the context in which it is employed. Some philosophers have criticized this idea on the ground that emotions are primary normative kinds and they can be given univocal descriptive analysis.

Even if Griffiths's claims are accepted in their entirety, all that can be inferred is that emotion is not a *natural* kind. This does not prevent us from scientifically examining it the same way as the concept *Vitamin* is not a natural kind but we can still investigate vitamins individually or in groups. What it suggests is that we should not expect a single theory to cover all aspects of emotion.

Barrett [7] argues that the assumption that certain emotions are given to us by nature and exist independent of our perception of them is wrong. This was based on an analysis of empirical findings in emotions research. She proposes instead that our perceptual processes lead us to aggregate emotional processing into categories that do not necessarily reveal the causal structure of the emotional processing. This argument seems to be of importance when discussing basic discrete emotions (section 2.2). It hardly affects continuous models of emotions that consider emotions as points in multidimensional affect space (section 2.2).

There are already many theories that try to explain emotions in psychology. James-Lange theory and its derivatives state that a changed situation leads to a changed bodily state and that emotion is simply the perception of this bodily change. In this model causation goes from the bodily state to the emotion and not the other way around (Fig. 2(a)). For example when seeing a fearful situation, specific body changes happen that the brain interprets as *feeling* fear. This theory and similar ones are supported by experiments in which emotion can be induced by modifying the bodily state.

At the other end of the spectrum Cannon-Bard theory claims that the causation goes from emotion to bodily condition. This means that the emotions arise first and *cause* the bodily state (Fig. 2(b)).

The Two-Factor theory attempts to combine the views of these conflicting theories and is based on experimental evidence that the emotional response of subjects is determined by two factors, the Bodily State in which they are in and the Perceived Context. In this case the causal relation becomes more complicated as there are two causes generating the emotion (Fig. 2(c)). A famous supporting experiment for this theory was done in [8]. The subjects were divided into two groups. The main group was injected with adrenaline while the control group was injected with a placebo. Every participant in the

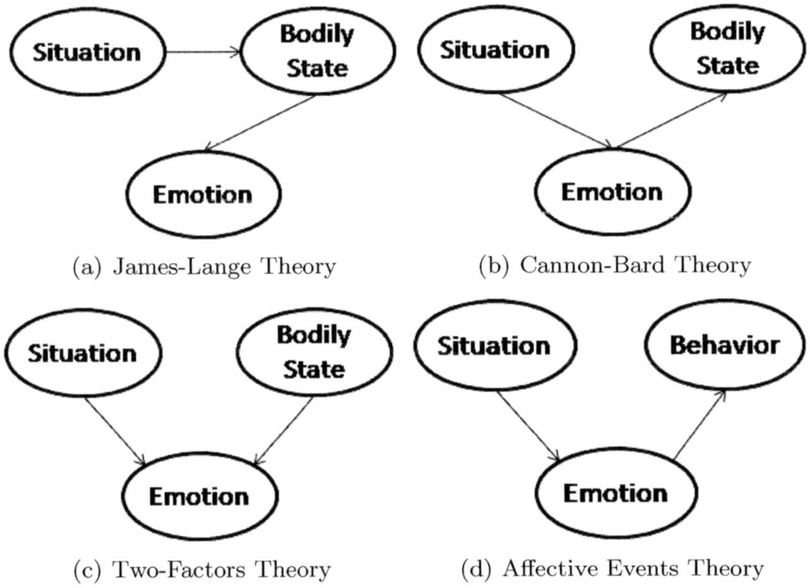

Fig. 2. Various models of the causal relation between emotion, bodily state, situation and behavior

main group was then put into a room with a confederate who shows either anger or amusement. The participants who were told that the confederate took the same injection behaved and reported feeling similar to the confederate. That is either angry or amused. Even though the physiological condition was the same due to the adrenaline injection, the emotion and behavior of the participant did depend on the situation.

The Affective-Event Theory developed by Howard Weiss tries to avoid any relation between bodily state and emotion. This is done by presenting the emotion generation within a communication framework. This theory suggests that the emotions are influenced and caused by events. In turn, these emotions influence attitudes and behaviors (Fig 2(d)).

In this chapter we use the causal theory shown in Fig. 3 to represent the relationship between emotion and its causes and effects. The main difference between this theory and the models shown in Fig. 2 is that we distinguish between two levels of emotion. The first level is called *affect* and represents mainly unconscious synchronization between bodily and cognitive components that trigger the conscious feeling of *feelings*. At the same time partially causes the behavior of the agent (which may be a human for example). The second level is what we call *feeling*. This represents mainly conscious feelings like anger, amusement, etc. For simplicity we have omitted the node for *rationalization*. This has a causal feedback loop with affect and causal links from the perceived situation and to the resulting behavior. Another minor difference between Fig. 3 and the models in Fig. 2 is that we explicitly

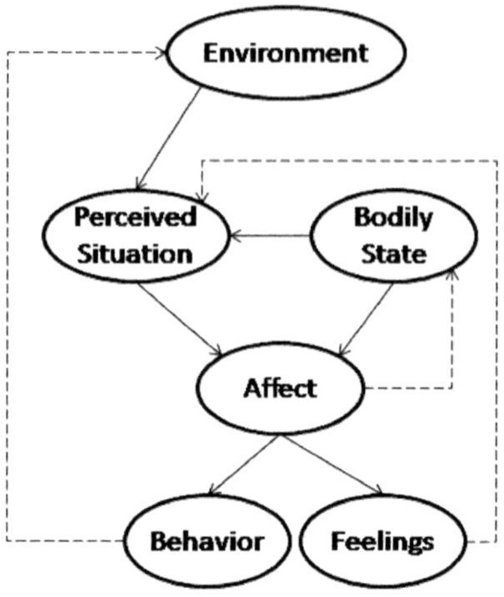

Fig. 3. Relationship between Emotion, Context, Bodily State, and Behavior according to our model

represented the environment and its causal relation to the perceived situation and the effect of behavior on it.

This theory supports the experimental evidence made in favor of the four theories mentioned earlier. This is done in a consistent way as all of the causal links found in the four theories (Fig. 2) can be found in this theory.

This theory also explains *emotional episodes* which are defined and widely used by the affective event theory. Emotional episodes are series of emotional states extended over time and organized around an underlying theme. In this theory *emotional episodes* arise because of the causal cycle: $Feeling \rightarrow Perceived\ Situatoin \rightarrow Affect$. This theory is also compatible with the component-process model [9] which describes emotion as the process by which low level cognitive appraisals triggers bodily reactions and actions. Here our *affect* nodes identifies with *emotion* as described by this cognitive model.

2.2 Models of High Level Emotions (Feelings)

Most of the research done in computational emotion modelling was focusing on the high level *emotions* represented by the *feelings* in Fig. 3. There are many models proposed for human emotions at this level.

In general there are two approaches used to model emotions either discrete basic emotions or continuous emotional spaces.

The first approach employs a set of discrete basic emotions. Ortony and Turner collected some of the most influential basic emotion categorizations according to the theorist as shown in Table 1. Fig. 4 show Plutchik's basic emotions in more details emphasizing that even in this discrete basic emotion model, each emotion has some strength. This is even though usually this strength is assigned a discrete rather than a continuous value as shown in

Table 1. Basic Discrete Emotions

Theorist	Basic Emotions
Plutchik	Acceptance, anger, anticipation, disgust, joy, fear, sadness, surprise
Arnold	Anger, aversion, courage, dejection, desire, despair, fear, hate, hope, love, sadness
Ekman, Friesen, and Ellsworth	Anger, disgust, fear, joy, sadness, surprise
Frijda	Desire, happiness, interest, surprise, wonder, sorrow
Gray	Rage and terror, anxiety, joy
Izard	Anger, contempt, disgust, distress, fear, guilt, interest, joy, shame, surprise
James	Fear, grief, love, rage
McDougall	Anger, disgust, elation, fear, subjection, tender-emotion, wonder
Mowrer	Pain, pleasure
Oatley and Johnson-Laird	Anger, disgust, anxiety, happiness, sadness
Panksepp	Expectancy, fear, rage, panic
Tomkins	Anger, interest, contempt, disgust, distress, fear, joy, shame, surprise
Watson	Fear, love, rage
Weiner and Graham	Happiness, sadness

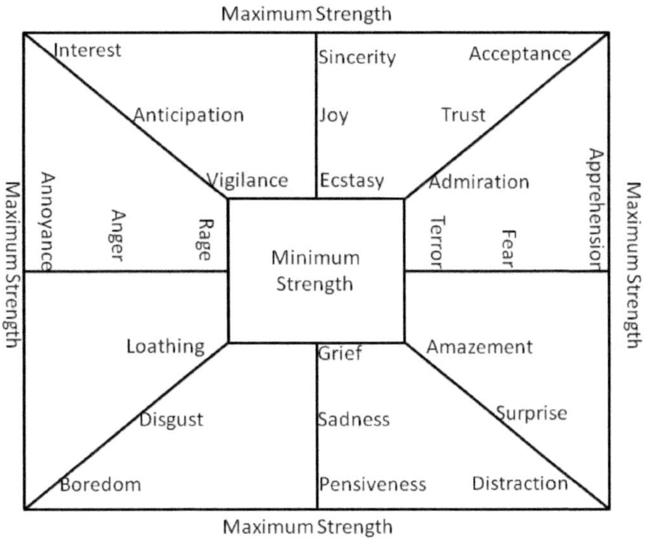

Fig. 4. Plutchik's basic emotions in details

Fig. 4. The conceptualization of emotions as discrete and independent has arisen mainly from research with animals. By selectively stimulating neural pathways and observing subsequent behaviors, or conversely by eliciting behaviors in highly constrained experimental circumstances and measuring neural activity, animal researchers have constructed taxonomies of the basic emotions and have proposed specific neural pathways associated with each putative basic emotion [10].

The main disadvantage of these discrete models is the difficulty in deciding what is really basic in these *basic* emotions. For example some models use only two basic emotions while others require up to eleven. It is not clear how an informed decision can be made about the *intrinsic* number of basic emotions required. Another problem of this approach is that it is hard to describe blends of emotions when the stimulus contains components which elicit more than one *basic* emotion. This discrete categorization of emotions assumes that emotions were evolutionary adaptations. In this case it is expected to have a limited number of these adaptations that had survival benefit to humans each served by its own independent neural pathway. Here Griffiths's claim may be considered to be valid as there is no reason to suppose that these individual separate adaptations have enough common properties to justify considering emotions a natural kind.

The second approach employs a continuous multidimensional space and each emotion is represented as a point in this space. The most commonly used model is the two dimensional arousal-valence model shown in Fig. 5. Valence represents how much pleasure the stimulus gives, with positive values

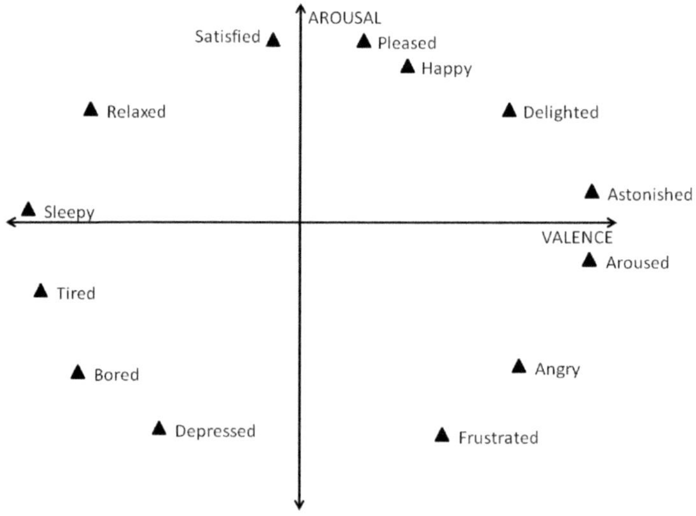

Fig. 5. 2D model of emotion

indicating pleasant and negative values indicating unpleasant stimulus. For example, happiness has a positive valence while distress has a negative valence. Arousal represents the activation level with higher activation levels assigned higher arousal values. For example agitation has high arousal value while relaxation has a low arousal value. The circumplex model proposed by Posner, Russell, and Peterson [10] asserts the following:

> All affective states arise from cognitive interpretations of core neural sensations that are the product of two independent neurophysiological systems. This model stands in contrast to theories of basic emotions, which posit that a discrete and independent neural system subserves every emotion.

Researchers have consistently reproduced the 2-D structure of the circumplex model using similarity ratings of facial expressions and emotion-denoting words Many of these findings have been replicated in a series of cross-cultural samples. Moreover, self-reports of affective states studied over various time frames, languages, and response formats have repeatedly yielded 2-D models of emotion. These findings all support the 2D model of emotion, but recent research found low consistency of physiological configurations associated with emotions in this model which indicates that ANS activation during emotions indicates the demand for action tendency as well as intrinsic emotion. This led Scholsberg to suggest a third [11] dimension of attention–rejection which is subsumed under the name stance leading to the three dimensional model depicted in Fig. 6.

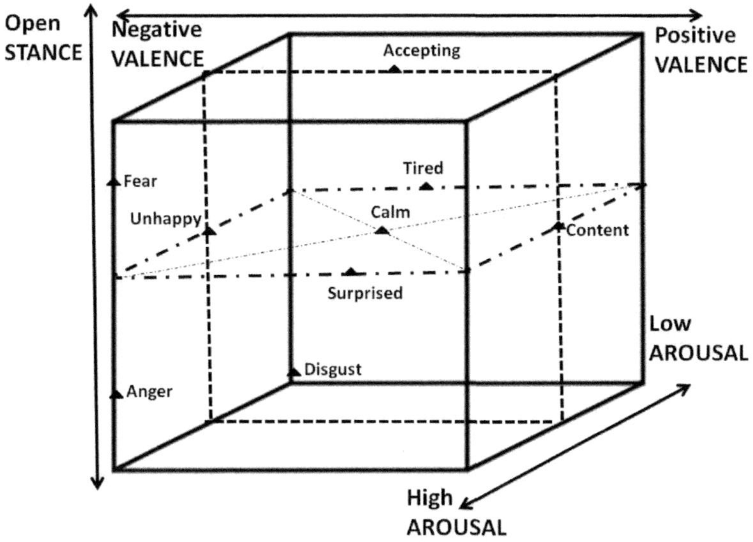

Fig. 6. 3D model of emotion

Adopting a theoretically based approach, Fontaine and others showed that, in the four languages (English, Dutch, French and Chinese), four dimensions are needed to satisfactorily represent similarities and differences in the meaning of words representing emotion. In order of importance, these dimensions were evaluation-pleasantness, potency-control, activation-arousal, and unpredictability. They were identified on the basis of the applicability of 144 features representing the six components of emotions. These are: (a) appraisals of events, (b) psychophysiological changes, (c) motor expressions, (d) action tendencies, (e) subjective experiences, and (f) emotion regulation [12].

Modelling of the low level *affect* state seems to be less well researched. Here, we focus on the low level *affect* part of emotion which is mostly subconscious. This complements the work in modelling high level emotions or feelings to provide a computational framework for the whole spectrum of emotions. Furthermore, we focus our attention on the *affect* during face to face interactions between agents. During these interactions, the affective state of one agent is dependent on the behavior and indirectly the state of the other agent. This can be seen in the adrenaline injection experiment presented earlier as the emotional response of the participant depended on how (s)he perceived the behavior of the confederate. Section 2.3 presents our model of this coupling and it forms the basis of this chapter.

2.3 Emotion during Interaction

Interaction between two human agents couples their internal state and their external behavior. Based on the model of emotion generation and effects presented in Fig. 3, we use a slightly modified model of emotion during interactions. This model is shown in Fig. 7. In this model the coupling between the two agents happens by using two mechanisms. The first mechanism is indirect and is caused by the changes the behavior of one agent causes in the common environment. This causes a change in the perceived situation of the other agent. The second coupling mechanism is the direct coupling that happens because the behavior of one agent is perceived by the other agent as a part of the perceived situation. That is perceived behavior in the figure. This direct coupling mechanism is affected by the interaction protocol assumed by each of the agents.

In this model, the interaction protocol assumed by an agent controls how this agent perceives the behavior of other agents. This in turn controls its emotional response to this behavior at the affect and feelings levels. For example, the act of tapping on someone's shoulder has a different *meaning* which causes different feelings depending on the situation, distribution of power, and other factors.

In this work we model the interaction protocol directly using dynamical systems and probabilistic networks.

Section 2.4 presents a detailed definition of the term *protocol* as used in this chapter. It provides also details of a computational model for representing, learning and adapting interaction protocols.

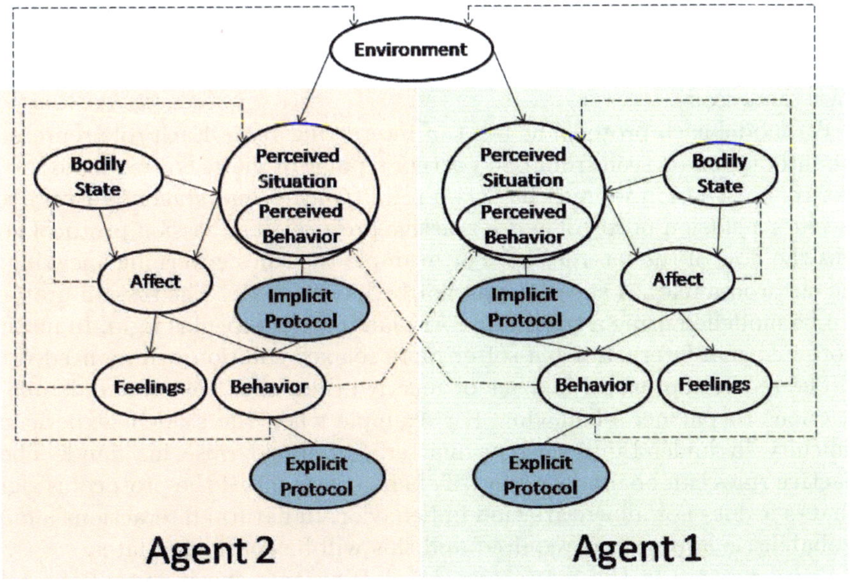

Fig. 7. Coupling between two agents during interaction

2.4 Interaction Protocol

Fig. 8 shows the model of the interaction protocol used in this chapter. We define the interaction protocol as a set of rules that govern the interaction. For example, in a classroom situation, there is usually an expected set of behaviors that the teacher and the students should execute to start the interaction (e.g. a greeting from the teacher and a limited set of postures and locations from the students), to keep it going (e.g. after a question from the

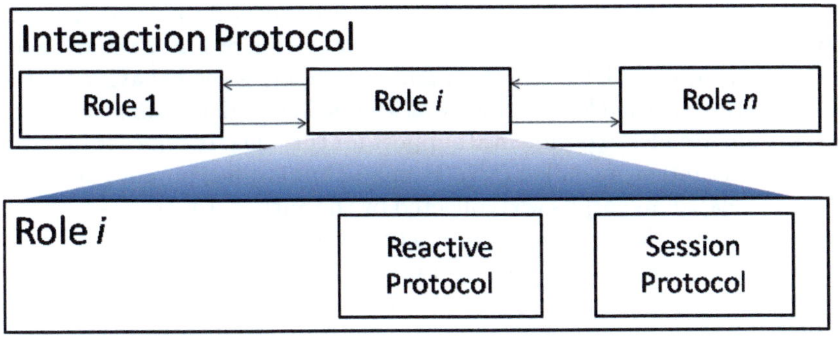

Fig. 8. Interaction Protocol as defined in this work

teacher, students are expected to react), and to finalize it. These behaviors need not be precisely defined in the protocol (e.g. locations can sit wherever they like in the room and the teacher may use different greeting styles to start the lesson).

We model each protocol as a set of interacting *roles*. Each role represents the interaction as seen from one partner's point of view. For example in the classroom situation we may have two roles (*teacher* and *student*). Every role specifies a session protocol and a reactive protocol. The session protocol governs the flow of the interaction. For example the rules governing speaking in the classroom like: "a speaker may not be interrupted". The session protocol can be modelled using a traditional AI plan if the protocol is rigid. In natural more informal interactions, a softer more relaxed solution will be needed.

The reactive protocol is a set of reactive rules that govern the details of reactions to partner's behavior. For example when the student experiences difficulty in understanding some matter, he should raise his hand. These reactive rules can be modelled by "if-then" constructs if the protocol is rigid. That is it does not allow variation in behavior. In natural interactions a more probabilistic approach is required and this will be considered later.

Every partner in the interaction has information specifying all the roles during the interaction. This is necessary for understanding the behavior of other partners who has other roles to play.For example, the student uses a model of how to behave as a teacher and uses this model to understand the teacher's behavior and may be to judge it.

The interaction protocols used by the other partners must be in harmony for the interaction to proceed seamlessly. For example if one of the students assumed the role *teacher* in the classroom situation, it is unlikely that the interaction will go smoothly.

2.5 Interaction Protocol and Affect

The main goal of this chapter is to provide a computational technique which can model both the affective state and the interaction protocols during face to face interactions. The question arises of why do we need to model both of these seemingly separate components. Theoretical consideration of Fig. 7 shows that the interaction protocol affects how each partner perceives the behavior of others. This in turn affects their low level state as well as their high level emotional state. This section provides experimental support for this idea. It comes from analyzing an interaction corpus called H^3R (Human-Human and Human-Robot Interaction Corpus). This corpus was collected by the authors to serve as a standard interaction corpus for evaluating interactive robots [1].

This corpus contains the results of 66 sessions conducted by 44 untrained subjects. Each subject acted either as a listener or an instructor. Instructors conducted 3 sessions each and interacted with two human listeners and one robot listener. The listeners conducted 2 sessions each. They interacted with two different human instructors. In each session, the instructor explains the

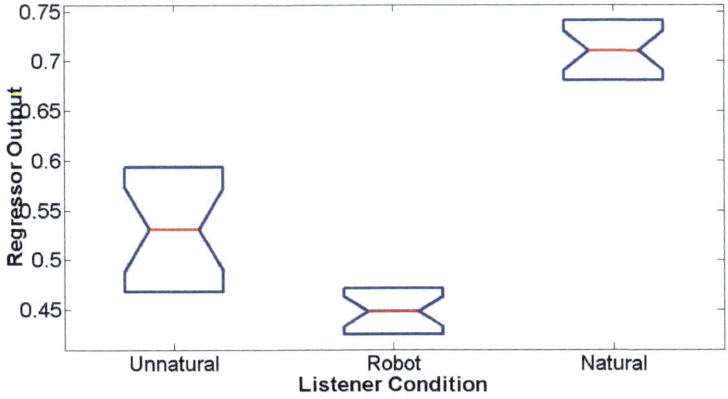

Fig. 9. Tree-Regressor's output for different listener conditions. It is clear that the regressor can distinguish the three conditions effectively based on physiological signals associated with emotional state.

assembly and disassembly of one of two devices to the listener. The listener was either a robot or a human. Human listeners interacted either naturally and attentively or unnaturally and exhibited distraction. We collected synchronized audio, video, motion, and physiological data of the listener and the instructor during every session. The details of this experiment are given in [1].

The first physiological signal we used was the conductance of the skin. Two channels related to skin conductance were used: Galvanic Skin Response [GSR] and Skin Conductance Level [SCL]. Mandryk and Inkpen [13] showed that GSR increases with the decrease of the task performance level or the increase in stress level. Shi et al. [14] have also shown that GSR is positively correlated to increased cognitive load when comparing the subjects response to a multimodal and a unimodal user interface. SCL is also correlated with affective arousal [15], frustration [16], and engagement [17]. The second sensor we used is that of a Blood Volume Pulse (BVP) sensor that measures heart activity.

The heart rate (HR) has also been used to differentiate between positive and negative emotions [18]. Heart rate variability (HRV) is also used extensively in human factors literature. It provides an indication of mental effort and stress in high stress environments [19]. Respiration is believed to be too slow for reflecting real time change in the internal state of humans. In [1] we showed that using appropriate processing it can be a reliable physiological differentiator between the response of an instructor to an attentive and an inattentive listener.

Our intention in this section is to compare the results obtained for a regressor trained on these physiological signals when the listener used a natural and unnatural interaction protocol. The result of this is shown in Fig. 9. As shown, the use of a natural interaction protocol is highly correlated with the emotional state of the partner. This shows experimentally that the interaction protocol and the affect are related in face to face interaction. This motivates us to develop artifacts that can provide natural interaction protocols in order to generate more desirable positive affective states in their users.

3 Approaches to Modeling Cognition and Emotion

We can, in general, distinguish between two approaches in modeling cognitive components including emotions. The first approach is the *behavior modeling* approach and the second is the *mechanism modeling* approach.

The purpose of the behavior modeling approach is to produce a final behavior maximally similar to the behavior of the modeled cognitive agent (usually human), disregarding both the processes whereby these behaviors are attained as well as the structures involved.

This approach is mainly one of engineering as it provides little information about the mechanisms underlying the modeled phenomena. Nevertheless, it can provide a valuable tool for practical applications. These models can also help theoretical investigation by separating the essential features of the mechanism or by providing the basis for empirical studies.

Most studies in emotion eliciting and display in embodied agents and robotics can be considered as behavioral modeling studies. They tend to utilize a pre-established model of the final emotional behavior as found by empirical studies and to replicate this model in the robot or agent. For example Brezeal [20] used the 3D model of emotion presented in Fig. 6 and the known connections between emotional state and facial expression. This was used to generate the behavior of a robotic head and then this behavior was compared with the known facial expressions of humans in similar situations.

The mechanism modeling approach is more ambitious and tries to model the hypothesized underlying mechanisms and structures responsible for the generation of modeled behavior. These studies can sometimes be less valuable as tools for practical applications but they can provide a grounded technique to compare, to refute, or to support scientific theories about the modeled cognitive component. At some level of abstraction mechanism models must rely on a behavior modeling approach for the implementation of their most basic components.

Mechanistic models need to be faithful to the input-output relations of the cognitive components they are modeling (exactly like behavior models) but moreover they need to be built on sound theoretical foundations that do not disagree with established facts or theories about the cognitive ability which is to be modeled.

An example of mechanism models is the work of Toda [21]. This work describes the behavior of the hypothesized Fungus Eaters in terms of urges and cost functions. That is internally versus externally.

Sometimes the distinction between behavior modeling and mechanism modeling is not very clear cut, specially when multiple competing theories of the mechanism exist. In this case it is very hard to develop a mechanism model that can be said to be good under all of these models. Simultaneously, some behavior models may provide new theories which explain the behavior resulting in deeper mechanism theory. Here we take the direction of mechanism modeling when dealing with emotions. We also explore both behavior modeling and mechanism modelling when dealing with interaction protocols. This will become clearer in later sections.

4 Modelling Affective State

According to the model shown in Fig. 3, the affective state acts as a mediator between perception and behavior that complements rational decision making. Research on how humans make decisions suggest a similar idea. Studies of decision-making by neurological patients who are unable to process emotional information normally suggests that people make judgments not only by evaluating the consequences and their probability of occurring, but also and even sometimes primarily at a gut or emotional level. Lesions in the ventromedial (which includes the orbitofrontal) sector of the prefrontal cortex interfere with the normal processing of *somatic* or emotional signals. At the same time sparing the most basic cognitive functions. Such damage leads to

Fig. 10. The robot used in the experiment

impairments in the decision-making process. This seriously compromises the quality of decisions made in daily life [22].

In [23], we detailed an experiment used for analyzing nonverbal interaction between untrained human subjects and a miniature nonhumanoid robot in a collaborative navigation task. The goal of this experiment was to analyze how nonverbal behaviors as gestures were used in situation and consequently finding signs of human adaptation to robot's behavior.

In this section we focus on the implementation of the robot presented in this study. It reflects directly our theory of affect mediated behavior as presented in Fig. 3. We, deliberately, avoid using the terms emotion and affect directly in this section in order to avoid confusion between the low level affect state and higher level *feelings* level. We use the word *mode* to represent any component of the affective state.

In this experiment the subject was required to guide the robot by means of hand gestures to follow a path projected on the ground. Along this path five types of virtual obstacles exist but cannot be seen by the subject. The robot cannot see the path but it can detect the obstacles when it is near to them. When facing an obstacle, the robot was supposed to give feedback to the robot using either verbal or nonverbal behavior.

The robot used in this experiment was a miniature e-puck [24] robot designed to study nonverbal communication between subjects and the robot in a collaborative navigation situation. The goal of this robot was to balance its internal drive to avoid various kinds of obstacles and objects during navigation with its other internal drive to follow the instructions of the operator. This operator cannot see the obstacles. The final responsibility of the robot was to give understandable feedback to help the operator correct her navigational commands.

The main feature of the control software in this experiment is the use of Mode Mediated Mapping. This means that the behavioral subsystem controlling the actuators is only connected to the perceptual subsystem representing the sensory information through a set of processes called *modes*. These modes represent the low level *affective* state (Fig. 3). Each mode continuously uses the sensory information to update a real value representing its controlled variable. The behavioral subsystem then uses these modes for decision making rather than the raw sensory information. Those modes constitute a representation of the internal state of the robot that governs its behavior. The components of the control system is shown in Fig. 11. The system consists of five main parts:

Localization System. This subsystem uses a combination of dead reckoning and vision based localization to localize the robot in the map within a circle of 2mm radius, and detects the direction of the robot with an average error of less than 2 degrees.

Perceptual Processes. This subsystem is responsible of detecting events that are important to the robot like an approaching obstacle, or a command from the GT software.

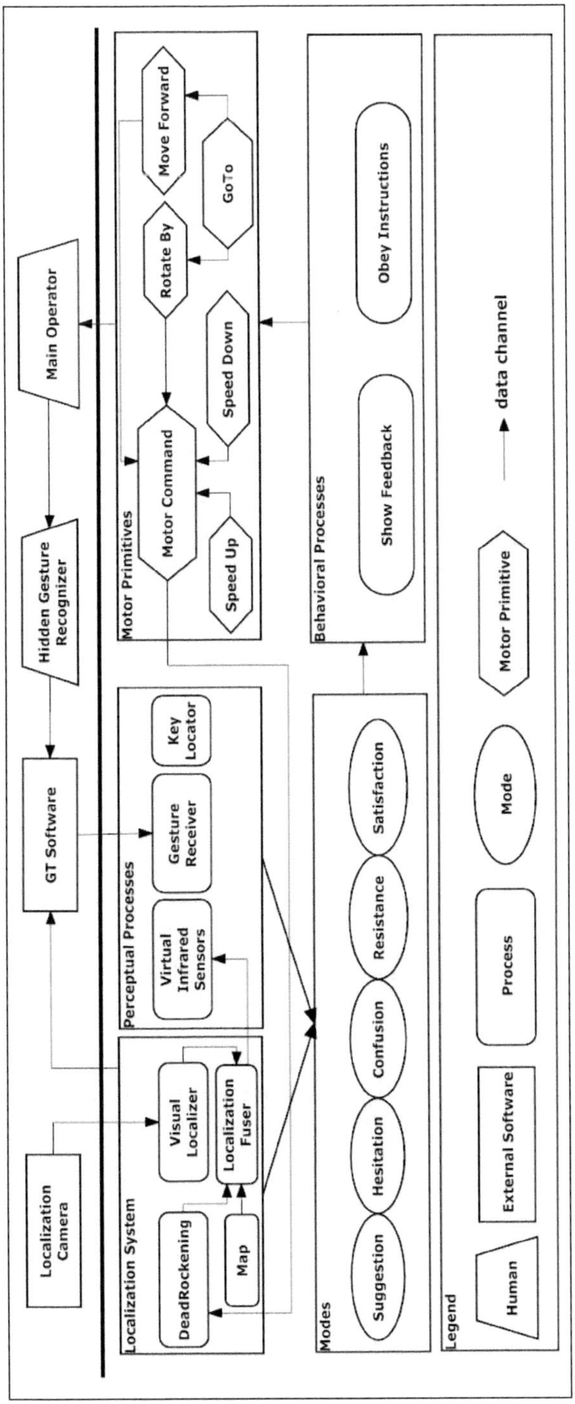

Fig. 11. The robot control software

Modes. A set of five modes each of which is represented by a real number in the range between 0 and 1. Those modes are continuously updated based on the following factors:

1. Sensed signals. For example if the robot did not receive a command for more than 15 seconds, its confusion mode increases.
2. Internal Evolution low. For example once the confusion mode reaches a value of over 0.5 it automatically increases with a rate of 0.05/second until a further external event such as a command from the user resets it lower than 0.6.

The five modes are combined as a single *mode vector* or an *affect state* used to guide the behavioral processes of the robot. This arrangement isolates the behavioral processes from noise in the input signals.

Behavioral Processes. The robot is mainly guided by six behavioral processes. The first process which is called *Obey Instructions* causes the robot to follow the path using the commands from the main operator. The five other processes are dedicated to give feedback signals once the modes of the robot approach some preset nominal values. These processes are responsible of closing the interaction loop between the operator and the robot. Table 2 gives the nominal values for the mode vector that triggers various feedback signals. Those modes were selected based on their performance in an exploratory study reported in [25].

Table 2. Nominal Mode Values for Various Feedback Signals

Feedback Signal	Meaning	Confusion	Suggestion	Resistance	Hesitation	Satisfaction
Signal 1	What should I do now?	1.0	0	0	0.5	0
Signal 2	I cannot pass in this direction.	0.1	0.8	0	0.8	0
Signal 3	What about this direction?	0	1.0	0	0	0
Signal 4	It will be too slow.	0	0	1.0	1.0	0
Signal 5	I found the key	0	0.9	0	0	1.0

Motor Primitives. This subsystem stores a set of low level motor primitives that are used by the behavioral subsystem to move the robot in the environment.

The specific technique used in the experiment to trigger feedback messaging (using modes as a mediator between the perceptual and behavioral subsystems) is more effective than hard-coding the triggering conditions without the affective mediation because:

1. This technique has the potential to reduce the effect of noise on the behavioral subsystem by buffering the signals through the modes. Fig. 12-A shows the gestures of the main operator while trying to stop the robot, rotate it counterclockwise then stop it again along with the actual gesture commands received by the robot. Fig. 12-B shows the angle taken

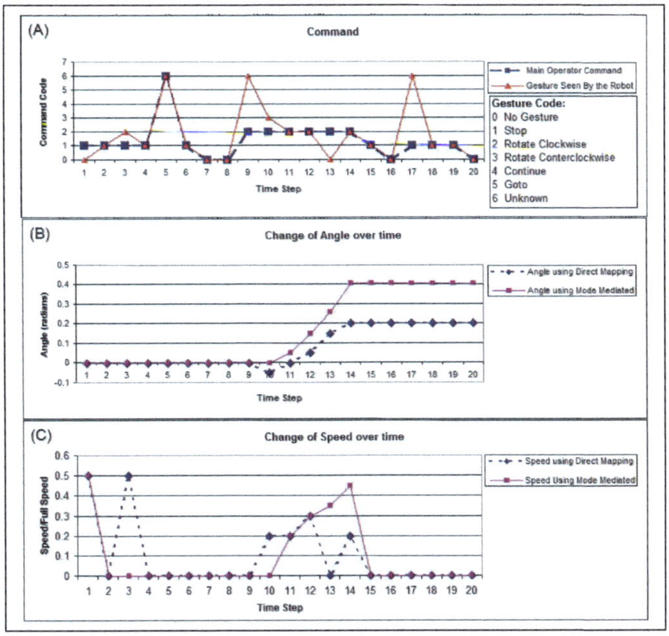

Fig. 12. Effect of using Mode Mediated Mapping on reducing the noise caused by gesture recognition

by the robot when assuming zero initial angle. From the figure it is clear that the mode mediated approach proposed here is more efficient. This is because the erroneous gestures introduced by the gesture recognition system causes the *confusion* mode to increase. This protects the robot from responding to such commands. Fig. 12-C shows the speed of the rotation of the robot. The mode mediated mapping approach improved the performance by causing the robot speed reflecting the main operator's intention better.

Fig. 13 shows another example of the use of mode mediated mapping when used to reduce the effects of environmental noise. As in Fig. 13-A, the noise in the distance between the robot and an obstacle is smoothed out in the distance between the current mode vector and the nominal value for giving suggestion feedback. This has two effects. Firstly, the robot did not give very short feedback attempts as in the direct mapping case shown in Fig. 13-B. Secondly, the robot gave the feedback long enough for the user to understand it and to avoid a collision.

2. It easier to extend the system. For example a special hazard detector process added later to the robot can modify the *hesitation* emotion in order to trigger the appropriate feedback without modifying its triggering preconditions.

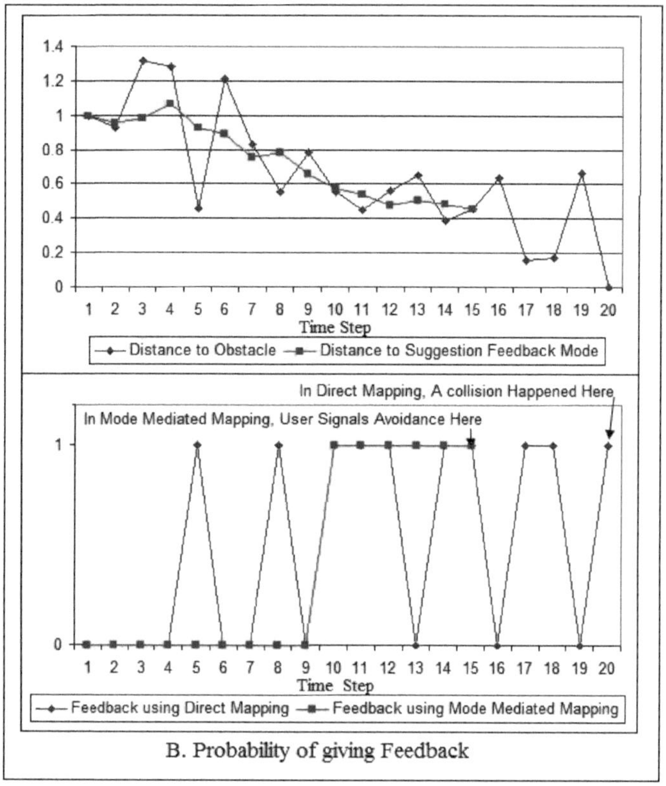

Fig. 13. Effect of using Mode Mediated Mapping on the navigation of the robot

5 Modelling Interaction Protocol

The model of interaction protocol depicted in Fig. 8 can be computationally implemented in different ways. A simple behavioral modeling approach to the problem is to consider how humans behave during interactions and to directly simulate this behavior. We took this direction in section 5.1.

Another mechanism modeling approach is to search neuroscience and cognitive science literature for information on how humans implement this kind of model. The protocol represents both the behavior of the agent and the behaviors expected from other agents. It combines behavior generation and an understanding of other partner's intentions and behavior. This later understanding is called a *theory of mind* and was the subject of much research in cognitive science and neuroscience. This is the direction followed in section 5.2. The two approaches will be compared in section 5.4.

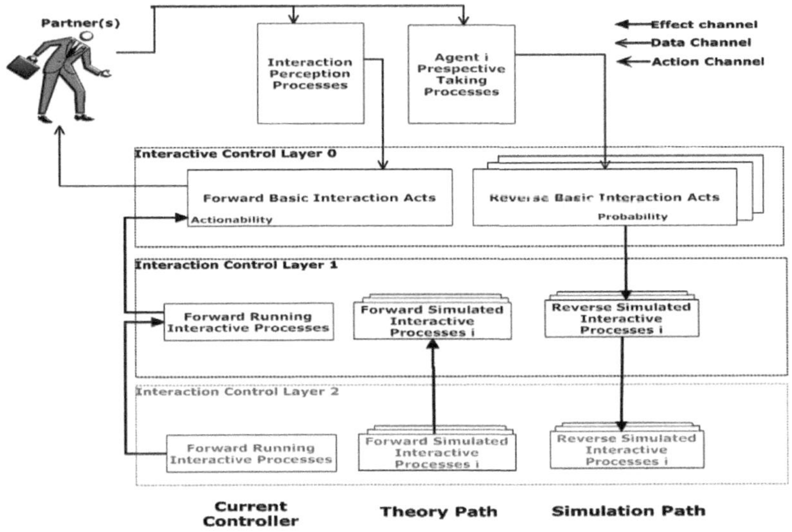

Fig. 14. Modelling Interaction Protocol using a hierarchy of probabilistic networks

5.1 Behavior Model Based on Human-Human Interaction Studies

During their interactions, humans utilize both verbal and non-verbal communication channels [26]. To manage these channels, the agent needs to have a variety of skills and abilities. These include dialog management, the synchronization of verbal and nonverbal behavior, and the efficient utilization of society-dependent spontaneous nonverbal behavior patterns. These skills are managed in humans by using both conscious and unconscious processes of a wide variety of computational loads. One of the most important skills in close encounters is using natural gaze patterns [26].

Research in gaze control for humanoid robots had focused mainly on the problem of detecting the gaze pattern of the partner and using simple heuristics to drive the gaze controller of the robot using this knowledge.Atienza and Zelinsky [27] used a stereo vision system to track the direction where a person is gazing. When the robot detects steady gaze at some object, it picks it up. A steady gaze was detected by using a heuristic about the distribution of gaze. Kuno et al. [28] proposed a method of two-way eye contact for human-robot communication. When a human wants to start communication with a robot, she watches the robot. When it finds a human looking at it, the robot turns to her, changing its facial expressions to let her know its awareness of her gaze. Seemann et al. [29] built a stereo vision based system to detect and track partner's gaze in real time. The main limitation of these systems and similar ones is the hand-coding of the gaze behavior of the robot based on heuristics. It has been shown in [26] that gaze behavior in close

Table 3. Comparison Between the Simulated and Natural Behavior for the gaze controller. All values are measured as percentage of the interaction time.

Item	Statistic	Ground Truth	Fixed Structure
Mutual Gaze	Mean	28.15%*	26.72%
	Std.Dev.	5.73%	1.67%
Gaze Toward Instructor	Mean	32.98%*	27.80%
	Std.Dev.	8.23%	4.28%
Mutual Attention	Mean	57.23%*	60.62%
	Std.Dev.	5.13%	4.97%

* According to [1].

interactions depends on many factors. It is very cumbersome to hand-code all the heuristics required to simulate human behavior.

Sinder et al. [30] studied the gaze behavior of a robot during conversations with humans. Hoffman et al. [31] designed a probabilistic system to model gaze imitation and shared attention. The rules for guiding the gaze behavior of the robot were also hand-coded in both of these systems and the first system used verbal information to guide the gaze controller.

The gaze controller implemented in this section was inspired by research on human nonverbal behavior during close encounters. Four reactive motor plans were designed that encapsulate the possible interaction actions that the robot can generate. These are looking around, following the human face, following the salient object in the environment, and looking at the same place as the human. These motor plans where implemented as simple state machines. The sufficiency of those motor plans was based on the fact that in the current situation the robot simply has no other place to which to look. The necessity was confirmed empirically by the fact that the three behavioral processes mentioned earlier were needed to adjust the activation level of all of these motor plans.

To design the behavioral level integration processes of the system we investigated existing research work on human-human nonverbal behavior in close encounters. A common mechanism for control of these behaviors – including proximities and body alignment – is believed to be the *Approach-Avoidance* mechanism suggested in [26]. This deals with managing spatial distance between interactors. The mechanism consists of two processes, one of them is pulling the agent to its interactor and the other pushes it away. The final distance between the interactors is determined by the relative strength of each of these two processes at any given point of time.

Since most explanation situations involve objects as well as the two interactors, a third process is needed to generate *mutual attention* to the objects of interest. The behavioral level integration layer of this fixed structure controller uses three processes.

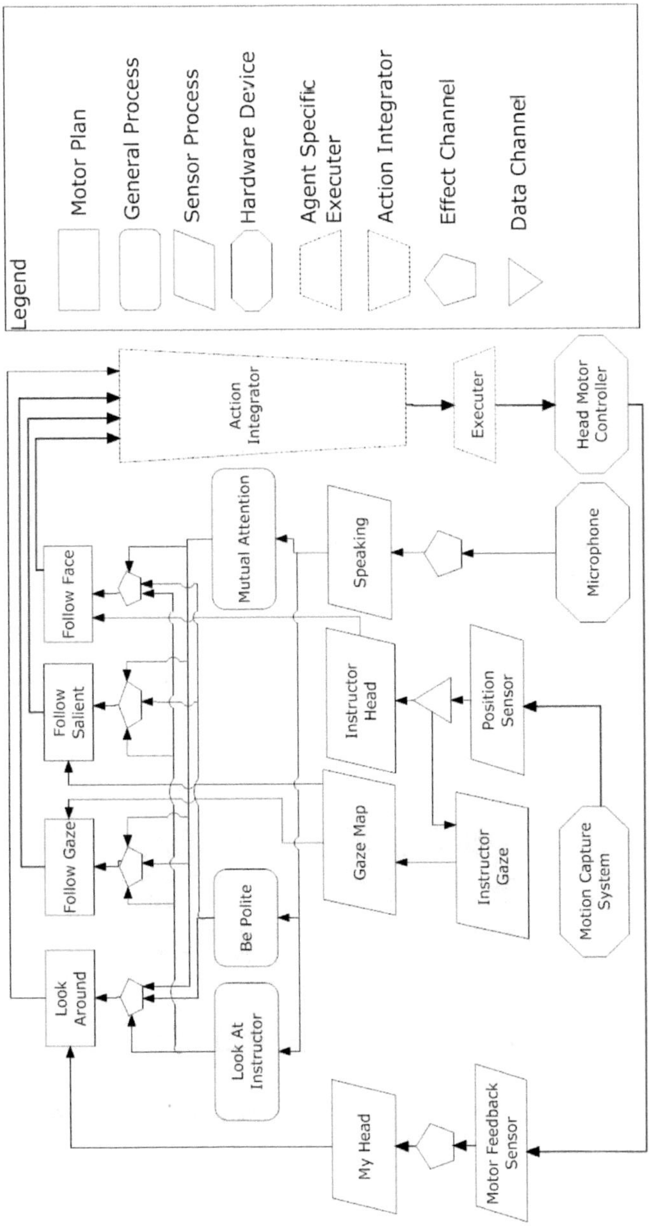

Fig. 15. The fixed structure gaze controller

1. *Look-At-Instructor*: This process is responsible for generating an attractive virtual force that pulls the robot's head in the direction of the human face.
2. *Be-Polite*: This process works counter to the *Look-At-Instructor* process. It provides the second force in the aforementioned *Approach-Avoidance* mechanism.
3. *Mutual-Attention*: This process seeks to make the robot look at the most salient object in the environment at which the instructor is looking at any given time.

Fig. 15 shows the complete design of this gaze controller. Because the number of processes in this controller is fixed during runtime, we call it the *Fixed Structure* gaze controller.

Table 3 shows that the behavior of the gaze controller (fourth column) is similar to the known behavior in the human-human case (third column) for the average times of the three behaviors. The standard deviation in all cases is less than 13% of the mean value which predicts robust operation in real world situations. These results suggest that the proposed approach is at a minimum be applicable to implement natural gaze control.

To find the similarity between $H(t)$, $G_f(t)$ and $G_d(t)$ we used Levenshtein distance [32]. In this experiment three baseline gaze controllers were used as control conditions: *Random*, *Follow* , and *Stare*. *Random* controller randomly selects a target and looks at it. *Follow* controller always follows the gaze of the instructor. *Stare* controller always looks at the instructor. Fig. 16 shows the behavior of the two proposed gaze controllers and control gaze controllers during one session of 5.6 minutes. The edit distance between every gaze controller's behavior and the human listener's behavior is also shown.

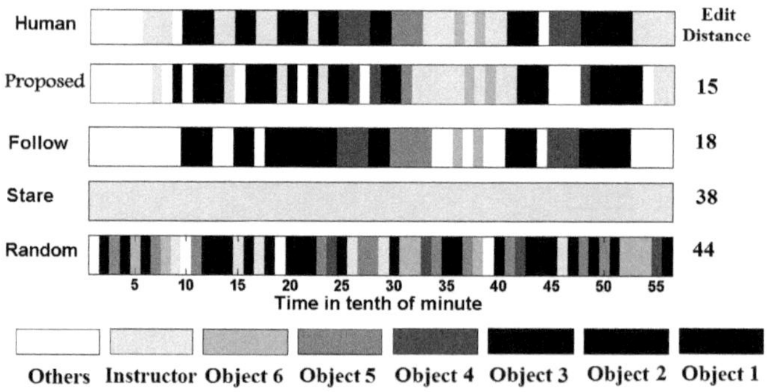

Fig. 16. The gaze target of the human listener, the proposed gaze controller and control gaze controllers with the edit distance between every gaze controller's behavior and the human listener's behavior. Object 1 to 6 represent objects in the environment related to the explanation situation.

The analysis reported in this section targeted comparing the external behavior of the controller with actual human behavior. This was done in accordance with the goals of behavioral modeling. In section 5.4 we compare the subjective evaluation to the behavior of this controller and the mechanism model presented in section 5.2.

5.2 Mechanism Model Based on Simulation Theory

To understand the intentions of other people, humans develop a theory of mind that tries to understand the actions of interacting partners in a goal directed manner. Failure to develop this theory of mind is hypothesized to be a major factor in developing autism and other interaction disorders [33]. Two major theories are competing to explain how humans learn and encode the theory of mind namely the *theory of theory* and the *theory of simulations* [33]. The *theory of theory* hypothesizes that a separate recognition mechanism is available that can decode the partner's behavior while the *theory of simulation* suggests that the same neuronal circuitry is used for both generation of actions and recognition of those actions when performed by others [34]. The discovery of mirror neurons in the F5 premotor cortex area of monkeys [34] and recent evidence of their existence in humans [35] support the theory of simulation although the possibility of the existence of other separate recognition mechanism cannot be ruled out.

In [36], we proposed the computational model shown in Fig. 14. This model is inspired by the *theory of simulation* presented earlier. Information flows in three interacting paths. Every basic behavior related to the interaction is encoded as what we call a *Basic Interactive Act*. Each BIA can be activated in two directions. When activated in the forward direction the BIA executes its behavior (e.g. nods, aligns body orientation with partner etc). When activated in the reverse direction, it indicates that a partner is executing the behavior.

This architecture is called L_iEICA. The main insights used in the design of L_iEICA are the following:

1. Generation of behavior in humans tends to employ bottom-up and top-down activation directions as well as reactive and deliberative processes. A robotic architecture capable of human-like natural interaction would possibly also include these combinations.
2. Knowing how to interact in a specific role in some interaction would entail at least some knowledge of how to interact in all the other roles. For example for a teacher to interact successfully with her students she must know at least a little about how it is to be a student because otherwise she cannot understand her students' behavior. This means that learning one role of the interaction should implies learning something about others. We will show that this is not the case in most machine learning approaches utilized in learning interaction structure. Our proposed architecture enables such combined learning easily.

3. Nonverbal interaction protocols and especially spontaneous ones are not specified at a single time resolution or abstraction level. They should be specified at various layers of abstractions corresponding to multiple time scales. The proposed level of specification can achieve this by using the idea of a multiple layers called *Interaction Control Layers.*
4. From the point of view of the cognitive processes, behaving in whatever role of the interaction should have similar if not the same computations inside the agent. This is a more involved view point that is based on both the *theory of simulation* in developmental psychology and mirror neurons in neuroscience. The proposed L_iEICA level of specification achieves this indistinguishability.

Social researchers discovered various levels of synchrony in natural interactions ranging from role switching during free conversations and slow turn taking during verbal interaction to the hypothesized gestural dance ([37]). To achieve natural interaction with humans, the agent needs to synchronize its behavior with the behavior of the human on a different time scale using different kinds of process ranging from deliberative role switching to reactive body alignment.

The L_iEICA system tries to achieve this by allowing the agent to discover how to synchronize its behavior with its partner(s) on appropriate timescales. The architecture is a layered control architecture consisting of multiple interaction control layers. Within each layer a set of interactive processes provide the competencies needed to synchronize the behavior of the agent with the behavior of its partner(s). It is based on a global role variable that specifies the role of the agent in the interaction process.

The goal of the system is then translated to learn the optimal parameter vectors for the interactive processes that achieve the required synchronization as specified by the behavior of the target partners which may be human beings.

The main parts of the architecture are:

Interaction Perception Processes (IPP). These are used to sense the actions of the other agents.

Perspective Taking Processes (PTP). For every interacting partner a set of Perspective Taking Processes are formed to provide a view of the interaction from the partner's point of view. Those processes generate the same kinds of signals that are generated by the agent's Interaction Perception Processes but assuming that the agent is in the position of the partner.

Forward Basic Interaction Acts (FBIA). These are the basic interactive acts that of which the agent is capable. In the current version those acts must be specified by the designer using arbitrary logic. These processes must use the simplest possible logic and should be deterministic to simplify the design of the Reverse Basic Interaction Acts explained next.

Reverse Basic Interaction Acts (RBIA). Every FIMP has a reverse version that detects the probability of its execution in the signals perceived by the IPPs or the PTPs. Those are the first steps in both the simulation and theory paths of the system. It allows the agent to represent the acts it perceives in the same vocabulary that is used to generate its own actions. The FBIAs and RBIAs constitute the first interaction control layer in the system. The rest of the interaction control layers can be learned by the agent.

Interactive Control Processes (ICP). Those constitute the higher interactive control layers. Every interactive control process consists of two twin processes. The forward process is responsible for adjusting the activation level of various processes in the lower layer based on the interaction protocol. They are, in the same time, used to simulate the partner. The reverse processes represent the theory the agent has about the partner and the protocol. It is related to the forward processes in the same way as RIMPs are related to FIMPs.

Shared Variables. Three globally shared variables are needed. Firstly, a variable called *Role* represents the agent's role during interaction (e.g. listener, instructor, etc). Secondly, a variable called *Age* represents the age of the agent which is the total time of interactions the agent has recognized or engaged in. A third variable called *Robust* is initialized for every partner and stores the aggregated difference between the theory and the simulation of this partner. This variable is used in conjunction with the age to determine its learning rate.

During interactions the processes of every layer are divided into two sets based on the role of the agent in the current interaction. The first set is the running interactive processes that represent the processes generating the actual behavior of the agent and runs in the forward direction. The second set is the simulated interactive processes that represent the other roles in the interaction (one set is instantiated for every other agent) and run in both the forward and reverse directions.

For simplification a two-agent interaction scenario (e.g. a listener-speaker scenario) will be considered in this section. Generalization to interactions that involve more than two agents is straightforward. In the beginning the role and age variables has to be set based on the task and current situation. Once those variables are determined the running interactive processes start driving the agent during the interaction. The perspective taking perceptual processes continuously translate the input stream into the partner's frame of reference, while the reverse basic interaction acts are measuring the most probable value of the actionability of various basic interaction acts of him/her/it. This is then fed to the reverse processes in the higher layers to generate the expected actionability of all the ICPs. This constitutes the theory about the intention of the other agent at different levels of detail based on the learned interaction structure. This is moving from bottom up in the interactive control layer hierarchy.

The forward direction of processes representing the partner is also executed at the whole hierarchy to generate the expected actionability of each of them according to the simulation of the partner. This is moving from the top down in the hierarchy. The difference between the theory and the simulation is used at every layer to drive the adaptation system only if the difference is higher than a threshold that depends on the age of the agent (Currently we use a threshold that increases linearly with the age) (see [36] for details). After adaptation mirror training is used to bring the reverse and forward processes of the simulated partner together. In all cases a weighted sum of the theory and the simulation results is used as the final partner actionability level for all processes and is utilized by the forward running processes to drive the agent.

5.3 Learning the Interaction Protocol

Fig. 17 presents a simplified version of L_iEICA components showing the development stage at which each of them is learned. The reverse processes are learned from forward ones using mirror training as described in [36]. The remaining processes to learn are then the forward basic interactive acts (FBIAs) and forward interactive control processes (FICPs). This section briefly describes the algorithms used to learn these processes from interaction records and adapting them during actual interactions. The details of these algorithms and their evaluation is given in [36] and [38].

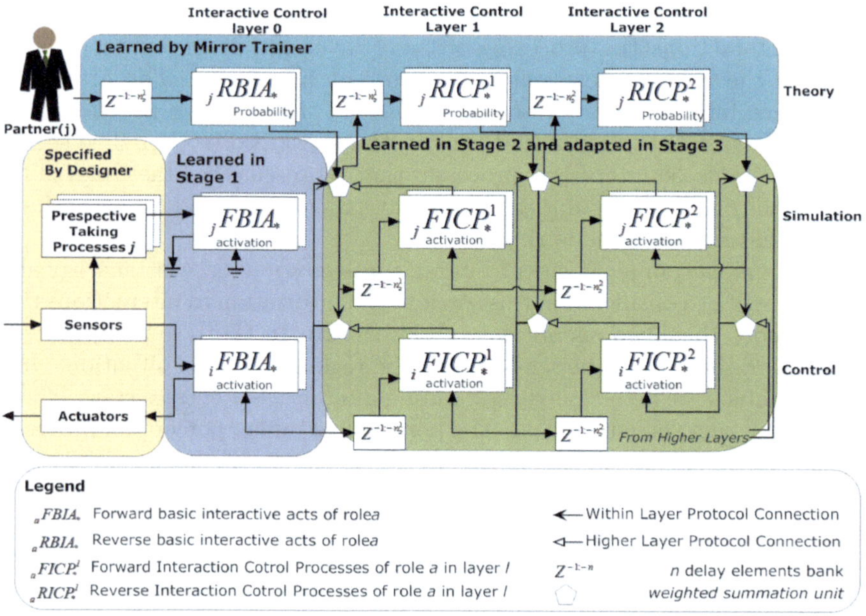

Fig. 17. L_iEICA components showing the learning algorithms used to develop each set

Stage 1: Autonomous Learning of Basic Interactive Acts (Interaction Bubbling). The first stage of development of the robot/agent aims at learning the forward basic interactive acts (FBIAs). This stage is called interaction bubbling to emphasize its relation with motor bubbling that allows new born babies to explore their motor abilities and learn the basic motor functions they can do. Similarly during interaction bubbling the robot (agent) learns how to use its sensors and actuators to achieve basic behaviors related to interacting with humans. The details of the algorithms used during this stage can be found in [38]. Here we provide an overview of the proposed technique.

The input to the learning mechanism are records of natural human-human or human-robot interactions. The robot first tries to discover recurrent patterns in the behavior of different actors (roles) in these interactions. The robot then associates a controller (dynamical system) with each of the discovered patterns capable of generating the required behavior (each such controller is a forward basic interactive act FBIA). Finally the mirror trainer is invoked to learn the reverse basic interactive act corresponding to each of the learned FBIAs.

The most critical step in this algorithm is the discovery of recurrent behavioral patterns. Given that the input to the robot is a multidimensional time series representing the behaviors of interacting agents, the problem can be coined as motif discovery from time series. There are many available techniques for solving this motif discovery problem. There is a common problem to all of these available algorithms which is their inability to utilize constraints or domain knowledge to speed up the discovery process which results in superlinear operation in all cases. Given that the length of the time series involved is usually high (e.g. hundred's of thousands or millions of time steps) in order to represent fast nonverbal behaviors like gaze shifts etc, a superlinear solution is too slow for our application. Also in this application the relations between the behaviors of interacting partners can be a useful clue for the probable locations of recurrent patterns (motifs) that are related to the interaction and can be useful for rejecting motifs that are not important for the interaction. Again available algorithms cannot utilize such relations to increase the accuracy and relevance of the discovered motifs. For these two reasons we defined the constrained motif discovery problem and provided three algorithms for solving it.

Stage 2: Autonomous Learning of Interaction Structure (Protocol Learning). Once the first interaction control layer consisting of the BIA's (both forward and reverse versions) is learned in stage one, the robot (agent) can start stage two of its development. The goal of this stage is to learn all the higher interaction control layers (both forward and reverse interaction control processes) using the same (or different) set of training examples used in the first stage. The algorithm used in this stage is called Interaction Structure Learning Algorithm (ISL) and is explained in details in [39].

Roughly speaking, the ISL algorithms builds ICPs from down up by adding new layers as needed to represent higher order (slower) synchronization protocols.

Once this stage is complete the robot can start interacting with human partners using the learned protocol and during these interactions in updates (adapts) the parameters of its ICPs and BIAs to better represent the interaction protocol. This online adaptation is the third stage of development that will be briefly explained in the following section.

Stage 3: Adaptation through Interaction. The final stage of development of the robot which continues to operate for the lifetime is the adaptation stage. During this stage the robot already have its whole architecture learned and it needs only to adjust the parameters of its BIAs and ICPs to best represent the interaction protocol. The algorithm used here is called the Interactive Adaptation Algorithm (IAA) and is presented in [36].

One tradeoff that the algorithm have to make is how much adaptive the robot should be. In general the robot will compare the behavior of its partner with the behavior it could have generated in case it was in its role and uses the difference to drive the adaptation algorithm (the details of how this is done is given in [36] and is not relevant to current discussion). Whenever a discrepancy is detected the robot has to decide whether or not it needs to revise its parameters. To control this decision we introduced the concept of *age* which is a global variable that specifies how old is the robot. If *age* is small, the robot tends to adapt more to the differences it finds between the behavior of its partners and its own protocol. If *age* is high, adaptation slows down or even stops after some threshold.

5.4 Comparing Mechanism and Behavior Models

In this section we provide the results of an experiment to compare the performance of L_iEICA based gaze controller with a carefully designed gaze controller that achieved human-like gaze behavior in [40].

In case of this controller the designer has to choose the required processes, their connections and parameter values. In case of L_iEICA, the designer needs only to specify the sensors, actuators and perspective taking processes and give the system human-human interaction records as training data. The robot then develops its own L_iEICA controller using the algorithms outlined in section 5.3.

The goal of this experiment is to compare the performance of of the behavioral approach presented in section 5.1 and the mechanism modeling approach presented in section 5.2. For this reason, in this experiment only stages 1 and 2 of the development were used and no adaptation was allowed. Subjective evaluations was used for comparison as in this work we are interested in the interplay between emotion and interaction protocol and behavioral evaluations cannot highlight the difference in this dimension.

Fig. 18. Snapshot of explanation scenario evaluation experiment

In this experiment we used an explanation scenario in which an instructor explains to the robot the operation of a device. The device is disassembled and put in a table between the listener and the instructor as shown in Fig. 18.

This experiment is designed as an internet poll and the behavior is evaluated using third-person subjective evaluation. This ensures high internal validity of the results at the expense of reducing external validity as this experiment cannot inform us whether the results to be found are generalizable to other situations.

Subjects where recruited from university students and staff. 38 subjects participated in the poll but the data of 3 of them was corrupted so we used only 35 subjects. The procedure of the session was as follows: Firstly, the subject is informed about the procedure using the following statement:

> This wizard will guide you in the steps needed to complete this survey.
> You will first answer a short questionnaire then you will watch two videos of the robot listening to an explanation about a device.
> After watching each video you will be asked to answer a few questions about the video you just watched. and that is ALL!!
> In every video there is an instructor explaining about a device to a humanoid robot.
> Notice that the speech of the instructor is not important and there are NO questions about it. You can even watch the videos without the sound.
> The procedure will take around 15 minutes (10 minutes of video and 5 minutes for answering).

Secondly, the subject answers the six questions to measure background information:

1. Age (ranged from 24 to 43 with an average of 31.16 years).
2. Gender (8 females and 30 males).

3. Experience in dealing with robots (ranged from *I never saw one before* to *I program robots routinely*).
4. Expectation of robot attention in a range from 1 to 7 (average was 4 with standard deviation of 1.376).
5. Expectation of robot's behavior naturalness in a range from 1 to 7 (average was 3.2 with standard deviation of 1.255).
6. Expectation of robot's behavior human-likeness in a range from 1 to 7 (average was 3.526 with standard deviation of 1.52).

After that the subject watches two videos: one showing the L_0EICA gaze controller and the other shows the controller developed using the mechanism explained in section 5.3. After each session the subject ranks the robot from 1 (worst) to 7 (best) in the following evaluation dimensions:

- Attention.
- Naturalness.
- Understanding instructor's explanation.
- Human-likeness.
- Instructor's comfort.
- Complexity of underlying algorithm.

Each subject is then asked to select his/her preferred robot from the two videos.

Both controllers achieved quit acceptable performance with an average rate of 4.94 for the L_iEICA controller and 4.49 for the L_0EICA gaze controller. The improvement achieved by L_iEICA is statistically significant according to two-samples ttest with p-value equal to 0.0243.

Fig. 19 shows graphically the difference between L_iEICA and L_0EICA controllers. From the figure, it is clear that L_iEICA outperforms L_0EICA

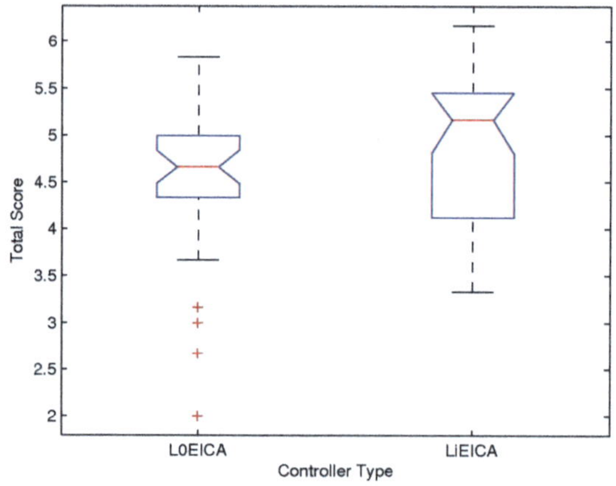

Fig. 19. Comparison between L_iEICA and L_0EICA gaze controllers in terms of total score

in average even though L_iEICA used only unsupervised algorithms that require no design decision other than the choice of the training set (and the motor primitives) while the L_0EICA controller was carefully designed using recent research results in human-human interaction and its parameters were adjusted using a Floating point GA to achieve maximum similarity to human behavior.

21 subjects selected L_iEICA controller as their preferred controller compared with 14 subjects for L_0EICA. This again supports the superiority of L_iEICA controller over L_0EICA controller in terms of total score.

6 Conclusion

In this chapter we argued that emotions need to be divided into two distinct levels. Low level emotions we call affect that represent unconscious internal state of the agent and high level emotions we call feelings that are usually represented either by discrete basic emotions or continuous emotion spaces.

We discussed a behavioral model of affect applied to a miniature robot and showed that this affect mediated implementation is both easier to extend and more faithful to current understanding of human cognition.

We then argued that interaction protocols are major cognitive components that need to be modeled in face to face interactions. Based on that we provided two approaches to model these interaction protocols, namely, the behavioral approach based on research in human-human interactions and the mechanism modeling approach based on theories from cognitive science and psychology. We compared two gaze controllers implemented using the aforementioned approaches and showed that the mechanism modeling approach provided more acceptable behavior according to participants' subjective evaluations and in the same time required no hard coding as the whole system is learned in an unsupervised way.

To realize intelligence in social context the three components of cognition we discussed (decision making, emotion, and interaction protocols) need to be implemented and interfaced correctly. This is we believe a fruitful direction of future research.

References

1. Mohammad, Y., Xu, Y., Matsumura, K., Nishida, T.: The h^3r explanation corpus:human-human and base human-robot interaction dataset. In: The fourth International Conference on Intelligent Sensors, Sensor Networks and Information Processing (ISSNIP 2008), December 2008, pp. 201–206 (2008)
2. Russell, S.J., Norvig, P.: Artificial intelligence: A modern approach. Prentice-Hall, Englewood (2003)
3. Searle, J.: Minds, brains and programs. Behavioral and Brain Sciences 3(3) (1980)

4. Yang, L., Yue, J., Zhang, X.: Hybrid control architecture in bio-mimetic robot, June 2008, pp. 5699–5703 (2008)
5. Ulam, P., Arkin, R.: Biasing behavioral activation with intent for an entertainment robot. Intelligent Service Robotics 1(3), 195–209 (2008)
6. Griffiths, P.: Is emotion a natural kind? In: Thinking about feeling, pp. 233–249. Oxford University Press, Oxford (2004)
7. Barrett, F.L.: Are emotions natural kinds? Perspectives on Psychological Science 1, 28–58 (2006)
8. Schachter, S., Singer, J.: Cognitive, social, and physiological determinants of emotional state. Psychological Review (69), 379–399 (1962)
9. Scherer, K.R.: Appraisal Considered as a Process of Multilevel Sequential Checking. In: Appraisal Processes in Emotion: Theory, Methods, Research, pp. 92–120. Oxford University Press, Oxford (2001)
10. Posner, J., Russell, J.A., Peterson, B.S.: The circumplex model of affect: an integrative approach to affective neuroscience, cognitive development, and psychopathology. Development and psychopathology (3), 715–734 (2005)
11. Scholsberg, H.: Three dimensions of emotions. Psychological Review (61), 81–88 (1954)
12. Fontaine, J.R., Scherer, K.R., Roesch, E.B., Ellsworth, P.C.: The world of emotions is not two-dimensional. Psychological Science 18(12), 1050–1057 (2007)
13. Mandryk, R.L., Inkpen, K.M.: Physiological indicators for the evaluation of co-located collaborative play. In: CSCW 2004: Proceedings of the 2004 ACM conference on Computer supported cooperative work, New York, NY, USA, pp. 102–111. ACM, New York (2004)
14. Shi, Y., Choi, E.H.C., Ruiz, N., Chen, F., Taib, R.: Galvanic skin respons (gsr) as an index of cognitive load. In: CHI 2007, April 2007, pp. 2651–2656 (2007)
15. Lang, P.J.: The emotion probe: Studies of motivation and attention. American Psychologiest 50(5), 285–372 (1995)
16. Lin, T., Hu, W., Omata, M., Imamiya, A.: Do physiological data relate to traditional usability indexes? In: OZCHI 2005 (November 2005)
17. Mower, E., Feil-Seifer, D.J., Mataric, M.J., Narayanan, S.: Investigating implicit cues for user state estimation in human-robot interaction using physiological measurements. In: 16th International Conference on Robot & Human Interactive Communication, August 2007, pp. 1125–1130 (2007)
18. Papillo, J.F., Shapiro, D.: The Cardiovascular System. In: Principles of Psychophysiology: Physical, Social, and Inferential Elements. Cambridge University Press, Cambridge (1990)
19. Rowe, D.W., Sibert, J., Irwin, D.: Heart rate variability: Indicator of user stateas an aid to human-computer interaction. In: Conference on Human Factors in Computing Systems, CHI 1998 (1998)
20. Breazeal, C.: Affective interaction between humans and robots. In: Kelemen, J., Sosík, P. (eds.) ECAL 2001. LNCS (LNAI), vol. 2159, pp. 582–591. Springer, Heidelberg (2001)
21. Toda, M.: Design of a fungus-eater. Behavioral Science 7, 164–183 (1962)
22. Bechara, A.: The role of emotion in decision-making: Evidence from neurological patients with orbitofrontal damage. Brain and Cognition 55(1), 30–40 (2004)

23. Mohammad, Y., Nishida, T.: Human adaptation to a miniature robot: Precursors of mutual adaptation. In: The 17th IEEE International Symposium on Robot and Human Interactive Communication, 2008. RO-MAN 2008, pp. 124–129 (2008)
24. EPFL: `http://www.e-puck.org`
25. Mohammad, Y.F.O., Nishida, T.: A new, hri inspired, view of intention. In: AAAI 2007 Workshop on Human Implications of Human-Robot Interactions, July 2007, pp. 21–27 (2007)
26. Argyle, M.: Bodily Communication, New Ed edition Routledge (2001)
27. Atienza, R., Zelinsky, E.: Intuitive human-robot interaction through active 3d gaze tracking. In: 11th Int. Symposium of Robotics Research (2003)
28. Kuno, Y., Sakurai, A., Miyauchi, D., Nakamura, A.: Two-way eye contact between humans and robots. In: ICMI 2004: Proceedings of the 6th international conference on Multimodal interfaces, New York, NY, USA, pp. 1–8. ACM, New York (2004)
29. Seemann, E., Nickel, K., Stiefelhagen, R.: Head pose estimation using stereo vision for human-robot interaction. In: Sixth IEEE International Conference on Automatic Face and Gesture Recognition, 2004, pp. 626–631 (2004)
30. Sidner, C.L., Kidd, C.D., Lee, C., Lesh, N.: Where to look: a study of human-robot engagement. In: IUI 2004: Proceedings of the 9th international conference on Intelligent user interfaces, New York, NY, USA, pp. 78–84. ACM, New York (2004)
31. Hoffman, M.W., Grimes, D.B., Shon, A.P., Rao, R.P.N.: A probabilistic model of gaze imitation and shared attention. Neural Netw. 19(3), 299–310 (2006)
32. Gusfield, D.: Algorithms on strings, trees, and sequences: computer science and computational biology. Cambridge University Press, Cambridge (1997)
33. Sabbagh, M.A.: Understanding orbitofrontal contributions to theory-of-mind reasoning: Implications for autism. Brain and Cognition (55), 209–219 (2004)
34. Murata, A., et al.: Object representation in the ventral premotor cortex (area f5) of the monkey. Journal of Neurophysiology 78, 2226–2230 (1997)
35. Oberman, L., et al.: Eeg evidence for mirror neuron activity during the observation of human and robot actions: Toward and analysis of the human qualities of interactive robots. Neurocomputing 70, 2194–2203 (2007)
36. Mohammad, Y., Nishida, T.: Toward combining autonomy and interactivity for social robots. AI & Society 24(1), 35–49
37. Kendon, A.: Movement coordination in social interaction: Some examples considered. Acta Pyschologica 32, 1–25 (1970)
38. Mohammad, Y., Nishida, T.: Constrained motif discovery. In: International Workshop on Data Mining and Statistical Science (DMSS 2008), September 2008, pp. 16–19 (2008)
39. Mohammad, Y., Nishida, T.: Toward agents that can learn nonverbal interactive behavior. In: IAPR Workshop on Cognitive Information Processing, pp. 164–169 (2008)
40. Mohammad, Y.F.O., Nishida, T.: A cross-platform robotic architecture for autonomous interactive robots. In: Nguyen, N.T., Borzemski, L., Grzech, A., Ali, M. (eds.) IEA/AIE 2008. LNCS (LNAI), vol. 5027, pp. 108–117. Springer, Heidelberg (2008)

Chapter 5
Emotional Model Based on Computational Intelligence for Partner Robots

Naoyuki Kubota and Shiho Wakisaka

Tokyo Metropolitan University,
kubota@tmu.ac.jp

Abstract. This chapter discusses the availability of computational intelligence to emotional model for partner robots. Recently, the need of human-friendly robots is increasing. Such a robot should have an emotional model in order to co-exist with people and to realize natural communication with people. In this chapter, we propose an emotional model based on location-dependent memory for partner robots using computational intelligence. First we discuss the functions of emotion in social interaction, and the availability of computational intelligence for the partner robots. Next, we propose an emotional model based on emotions, feelings, and mood. Furthermore, the episodic memory is related with the change of feelings. Therefore, we propose a map building method based on the emotional model. Next, we propose a behavior control method based on the emotional model. Finally, we show several experimental results of the proposed method, and discuss the availability of the emotional model based on location-dependent memory.

Keywords: Partner Robots, Emotional Model, Episode Memory, Interaction and Communication.

1 Introduction

Human expression used for natural communication is deeply related with emotional states. The relationship between human emotion and communication has long been discussed from the viewpoints of psychology, sociology, neurophysiology, and brain science [1-8]. Such discussions generally take one of two approaches; macroscopic and microscopic. Discussions of emotion in psychology tend to view the body macroscopically as a black box, and the internal processing in mind has been discussed. In discussions of the relationship between emotional expression and social behavior in sociology, Keltner and Kring discussed the functions of emotion in social interaction as (1) informative, (2) evocative, and (3) incentive [5]. In informative functions, emotional expression conveys information about senders themselves and about objects and events in the social environment. In evocative functions, an individual's emotional expression serves as a social affordance that evokes "prepared" responses in others. In incentive functions, an individual's emotional expression and experience may provide incentives for reinforcing another individual's social behavior within ongoing interactions.

T. Nishida et al. (Eds.): Modeling Machine Emotions for Realizing Intelligence, SIST 1, pp. 89–108.
springerlink.com

Emotion is discussed microscopically in terms of neurotransmitters and neuronal networks in brain science. The American Heritage Dictionary defines emotion as a mental state that arises subjectively rather than through conscious effort and is often accompanied by physiological changes [9]. The constructivist approach to emotion often involves modeling using mathematical and system models based on psychology and brain science theory. In short, research on emotion is very important in analyzing and understanding human physiology and sociology.

Natural communication based on emotion is widely discussed in research on human-friendly robots [10-12]. Human communication generally involves the perception of another's intention and feeling, which indicates that an emotional model is very helpful for human to understand the state of the robot. Emotions influence actions such as incentive functions of emotion. When a user feels happy, the human takes its corresponding pleasant action. Therefore, a human reaction is very useful to estimate the feeling of others. If robots had emotional models, perhaps they could communicate more smoothly and naturally with users. Robots thus need a human-like emotional mechanism. The robot's emotional model helps give meaning and value to the perceptual information, and indirectly directs decisions based on internal and external state. An emotional model has many potential uses, but an effective emotional model for robots should be constructed.

There are two main research streams on emotions. One is emotion modeling [13-14] to making parts of emotions based on the constructivist approach. Other work is to define emotions in multidimensional space. We also discussed the applicability of emotional models for robots based on constructivist approach [15,16]. In [17], an emotional model is applied to regulate the motion outputs of robots according to the time series of perceptual information. Next, we applied an emotional model for learning behaviors through interaction with people [18]. In this method, the learning rate is updated according to the feelings of the robot. As a result, a person can control what to learn in the learning indirectly by the control of the feeling of the robot using human natural communication such as praise and scolding. This method can be considered as one of reinforcement learning based on reward and penalty. Furthermore, we discussed the applicability of an emotional model through experiments for a partner robot to realize the incentive functions of emotion through the communication with people [19]. We thus discussed the emotional models for robots from various points of view. However, we have not yet discussed the applicability of long-term emotional memory between robots and people. In general, people sometimes have location-dependent emotion in episodic memory. For example, after a person had a traffic accident in a crossing, the person tries to avoid the crossing with a fear feeling. Some people go to a deserted park to deliberate deeply, and go to a dark seaside to forget their unpleasant memory based on their experiences. Therefore, we proposed map building for realizing location-dependent emotional memory [19].The performance of map building methods and behavior control methods is related with the technologies on computational intelligence. Therefore, we discuss the availability of computational intelligence to emotional model for partner robots in this chapter.

This chapter is organized as follows: Section 2 explains the background of computational intelligence for partner robots. Section 3 proposes an emotional model, map building method, and behavior control method for a partner robot. Section 4 shows experimental results using a partner robot and an emotional model based on location-dependent memory. Section 5 summarizes our findings, and discusses the future direction of this study.

2 Partner Robots Based on Computational Intelligence

2.1 Partner Robot; MOBiMac

The PC-type partner robot we developed, MOBiMac (Fig.1), [20] has two CPUs for PC and robotic behavior. The robot has two servomotors, eight ultrasonic sensors, a microphone, and a pan-tilt CCD camera. The partner robot should understand the meaning of user expressions (actions) in real environments (RE) and in virtual reality environments (VRE) (Fig. 2). MOBiMac can learn behavior in different environments. We used both RE and VRE constructed based on perceptual information from the environment [21,22]. In the VRE, a user interacts with robots by microphone and a CCD camera of putting in preparation for the partner robot's PC. We proposed the methods for gesture recognition, perception-based utterance, imitative learning, interactive learning with people, and others based on computational intelligence [23-25].

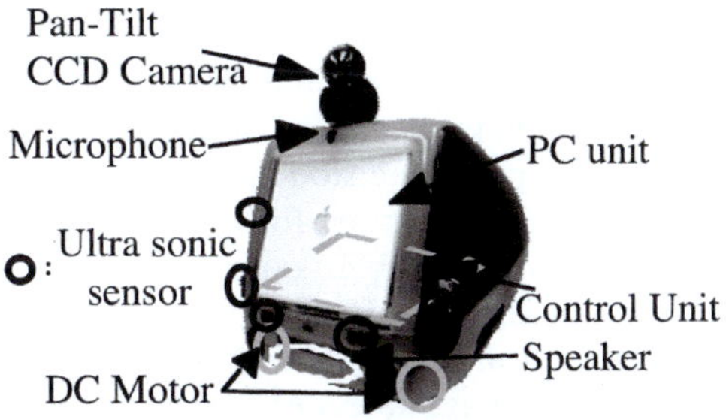

Fig. 1. "Partner" Robot MOBiMac

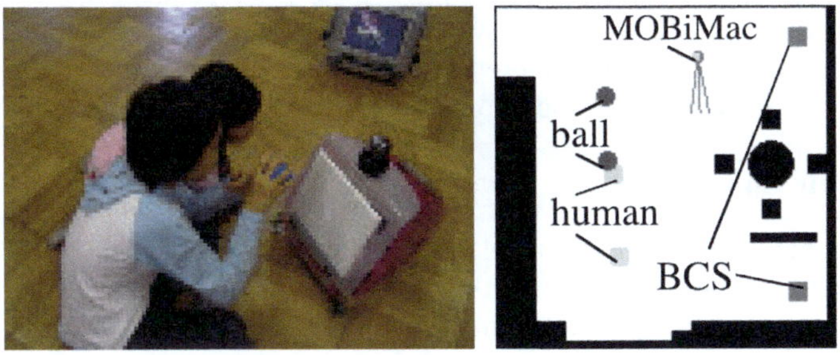

Fig. 2. Interaction of a partner robot with users in the RE and VRE

Upon detecting a potential user, the robot tries to communicate and interact with it, expressing internal and perceptual states and intentions by speech and behavior. The emotional model changes content of speech. Choosing among several behavior modes, the robot selects behavior based on the emotional states. The robot updates emotional states based on measured environmental information and user reactions extracted using image processing and speech recognition. The robot repeats interaction with a user based on the emotional model and user reaction.

2.2 Intelligent Technologies

Bezdek discussed intelligence from three levels: artificial, biological, and computational[26]. In the strictest sense, computational intelligence (CI) depends on numerical data and does not rely on explicit knowledge. Furthermore, Eberhart defined CI as a methodology involving computing. We also summarized CI as follows [27]. CI aims to construct intelligence from the viewpoints of biology, evolution, and self-organization. CI tries to construct intelligence by the bottom-up approach using internal description, while traditional artificial intelligence (AI) tries to construct intelligence by the top-down approach using external (explicit) description. However, these research fields use neural computing [28], fuzzy computing [29], and evolutionary computing [30] as intelligent techniques. Neural computing and fuzzy computing are based on the mechanism of human brain. While neural computing simulates physiological features of human brain, fuzzy computing simulates psychological features of human brain. Each technique is not complete for realizing all features of intelligence, and therefore, hybridized or combined methods have been proposed for building intelligent systems.

CI has been applied to perception, decision making, control, learning, and others of robots [31]. Neural computing is applied for the learning of perceptual systems and behavioral systems. Fuzzy computing is often used for human-like linguistic control and inferences. Evolutionary computing is used for the search in unknown environments or without environmental knowledge. Furthermore, evolutionary computing is used for path planning, trajectory planning, and task planning. Reinforcement learning is used for the learning in multi-stage decision making with reward and punishment (or success and failure) based on state-value functions [32]. In order to realize the episodic memory and location-dependent memory, we use a state-value function. Furthermore, we apply fuzzy control for multi-objective behavior coordination in unknown environments co-existing with people. In the following, we explain the background of behavior control and reinforcement learning.

2.3 Robotic Control

In a traditional approach, the control architecture for a mobile robot is designed by sequential functional decomposition. The first function is sensing and perception. In this step, sensing information from different sensors including vision, auditory, tactile, and others is integrated into a central representation. As internal processing, building or updating a model of the environment often called a world model, is performed. The next function is decision making. The robot tries to search and find the optimal action in the world model. Finally, the selected actions are executed in the real world. In this way, the sequential functional decomposition of a task leads to the sense-think-act cycle of the traditional information processing approach. The world model for collision avoidance and target tracing is often generated by the map building.

Symbolic representation, manipulation, and search in traditional AI are very useful and powerful in discrete space. Furthermore, if a precise world model is available, the robot can perform the motion planning and decision making suitable to the world model. However, representational and inferential frame problems arise from the complete generation of an internal world model and the complete search of actions in the internal world model. To avoid the difficulty of world modeling and map building, bounded rationality approaches have been proposed by trying to reduce the world modeling and search into a bounded amount of computation. Behavior-based robotics uses subsumption architecture as a quite different method from the traditional AI. In the subsumption architecture, the agent design is decomposed into objective-based behaviors such as obstacle avoiding, photo tracing, and map building. The behavior-based approach is considered as "non-information-processing-based". Basically, behaviors are designed using finite state machines, but neural networks and fuzzy systems have been also used for describing behavior rules. The subsumption architecture builds control architecture by incrementally adding task-achieving behaviors on top of each other. Higher layers can subsume lower layers. Higher-level layers build and rely on lower-level ones. The subsumption approach realizes the direct coupling between sensors and actuators, with only limited internal processing. A level of competence is the informal specification of a class of desired behaviors that the robot should be able to perform in the environments in which it will have to operate. The subsumption architecture is based on the key concepts of situatedness and embodiment, *i.e.*, the robots are situated in the world, and the robots have bodies and experience the world directly. The behavior-based robotics realizes a real-time control based on reactive motions in unknown or dynamic environments, but it is pointed out that the robot cannot perform sequential or complicated tasks. Therefore, in order to improve the capabilities of intelligent behaviors and human interactions, we propose behavior control methods based on reinforcement learning and emotional model.

2.4 Reinforcement Learning

A reinforcement learning method builds an agent learning the input-output pairs to maximize its expected utility only using reward or punishment from the environment (Fig.3). First of all, an agent perceives an environmental state (s_t) and reward signal (r_t), and takes action a_t at the discrete time t. Here, the dotted line indicates the indirect information flow from the environment, because the reward is perceived indirectly from the environment or given by the system. Next, the relationship between s_t and a_t is updated according to the reward signal (r_{t+1}) obtained at the next discrete time t+1. The detail is explained in the Section 3.2. Accordingly, the reinforcement learning does not use explicit teaching signals, but uses evaluative feedback obtained through the interaction with the environment. The tasks to be dealt with in the reinforcement learning are sequential decision problems based on four elements of a policy, a reward function, a value function, and optionally, a model of environment.

A policy is defined as a set of mapping rules from an observed state of the environment to an appropriate action. The policy can be represented as simple functions, lookup tables, neural networks, and fuzzy inference rules. After the agent takes an action according to the policy, the observed state of the environment is changed to other state. The agent receives a reward from the environment by taking an action. Basically, the terminal state is defined as the boundary condition. Each episode ends when the agent observes a terminal state. The study of the reinforcement learning is to build an intelligent agent or robot itself, because the agent of the reinforcement learning perceives the environment, makes decision according to the policy, takes action, and updates the value function. If the robot uses a geometric location as a state, the reinforcement learning is corresponding to the map building based on state-value functions. If the robot can refer to the value map, the robot can select suitable actions.

Behavioral learning methods based on reinforcement learning have been proposed by many researchers, but behavioral learning based on emotion-based reinforcement learning is not many. For example, Doya suggests that emotion enables a novel computational approach as a meta-learning system, and the studies on learning theory, autonomous agents, and neuromodulatory systems can be bound together [37].

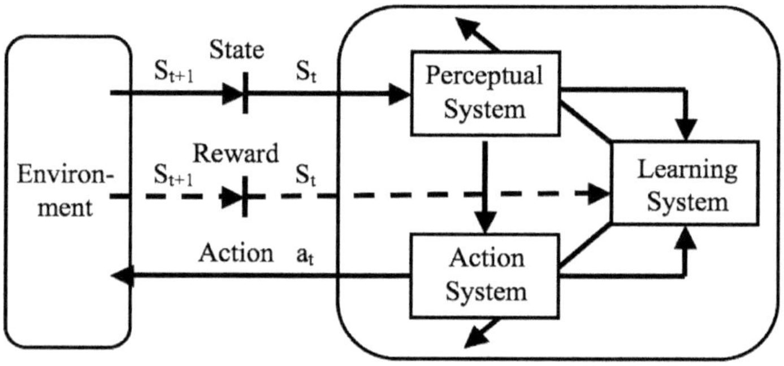

Fig. 3. A mechanism of reinforcement learning

2.5 The Role of Emotional Model in Computational Intelligence

CI can provide robots with intelligent capabilities similar with people. However, we must design evaluation functions for decision making and learning of robots. If the robot has some fundamental criteria, we don't need to design evaluation functions. The emotional model can be used for representing internal physical states and external environmental states as the relative and subjective criteria and values based on innate features or through experiences like humans.

The relationship between strategy and tactics has been discussed in various research fields. From the point of view, basically, CI can be considered as strategies in the decision making and learning, while the emotional model can be used for evaluating the state of decision making and learning in the tactics. In the following, we explain how to use an emotional model for partner robots based on computational intelligence.

3 Emotional Model for Partner Robots

3.1 Emotional Model Based on Perception

We apply the concepts of emotion, feeling, and mood based on a time scale, assuming that emotions change temporally based on the perceptual information on the internal state and the external environment. Emotion is considered an intense short-term mental state based on perceptual information and used as intermediate input from the perceptual system to the emotional model. Therefore, we assume that emotion changes dependent on specific perceptual information, and the robot independently converts perceptual information to emotional input based on predefined rules. For example, if the robot detects a user, input to the feeling of happiness is generated as an emotion. Each feeling is updated as the summation of emotions. The *i*th emotional input $u_{j,i}^{E}(t)$ is generated based on *j*th perceptual input as follows:

$$u_{j,i}^{E}(t) = z_{j,i} u_j^I(t) \tag{1}$$

where $z_{j,i}$ is the degree of contribution from the *j*th perceptual input to the *i*th emotion, $u_j^I(t)$ is 1 if the *j*th perceptual information is validated ($u_j^I(t)=\{0,1\}$), and N^I is the number of perceptual inputs.

The state of *i*th feeling $u_i^F(t)$ is updated by the emotional input from the viewpoint of the bottom-up construction, and by the top-down constraints from mood $u^M(t)$:

$$u_i^F(t) = \kappa u_i^F(t-1) + (1-\kappa)\sum_{j=1}^{N^I} u_{j,i}^E(t)$$

$$\kappa = \frac{\gamma^F}{1+u^M(t-1)} \tag{2}$$

where γ^F is the discount rate of feelings ($0<\gamma^F<1.0$).

Mood is defined as the long-term state updated by a change in feelings, and governs changes in feelings. Feeling is defined as a short-term state updated by a change in emotion (Fig.4). The state of mood is updated by the sum of feelings:

$$u^M(t) = \gamma^M u^M(t-1) + \frac{1-\gamma^M}{N^F}\sum_{i=1}^{N^F} u_i^F(t) \tag{3}$$

where γ^M is the discount rate and N^F is the number of feelings. Figure 5 shows an example of the emotional model over the time.

In the emotional model, we use parameters of happiness, fear, and anger as feelings ($N^F=3$), and the existence of users, balls, battery-chargers, and obstacles as perceptual input (Fig.2 (b)). We apply a simple spike response model, which is an artificial spiking neuron model [33] to generate perceptual image-processing input. Basically, a user quickly perceives an object or event having a strong corresponding stimulus, but requires time to perceive one having a correspondingly weak stimulus. This is strongly related to the summation of an external stimulus in the perceptual system. Internal state $h_i(t)$ is calculated as follows:

$$h_i(t) = h_i(t-1) + h_i^{ext}(t) + h_i^{ref}(t) \tag{4}$$

$h_i^{ext}(t)$ is input to the ith neuron from the external environment, and $h_i^{ref}(t)$ is used for neuron refractoriness. When the internal state of the ith neuron exceeds the predefined threshold, a pulse is output as follows:

$$u_j^I(t) = \begin{cases} 1 & \textit{if } h_i(t) \geq \theta_i \\ 0 & \textit{otherwise} \end{cases} \tag{5}$$

where θ_i is a firing threshold. When the neuron is fired, H is subtracted from refractoriness as follows:

$$h_i^{ref}(t) = \begin{cases} \gamma^{ref} h_i^{ref}(t-1) - H & \textit{if fired} \\ \gamma^{ref} h_i^{ref}(t-1) & \textit{otherwise} \end{cases} \tag{6}$$

where γ^{ref} is the refractoriness discount rate. The distance to the detected object is input $h_i^{ext}(t)$ from the external environment, so emotional input occurs as the robot approaches objects.

Speech recognition is conducted for human interaction, and the term acquired by speech recognition is categorized into negative and positive words corresponding to factors causing negative and positive emotions directly used as perceptual input $u_{i,j}^E(t)$. As a result, the states of feelings are updated by the meaning of recognized words.

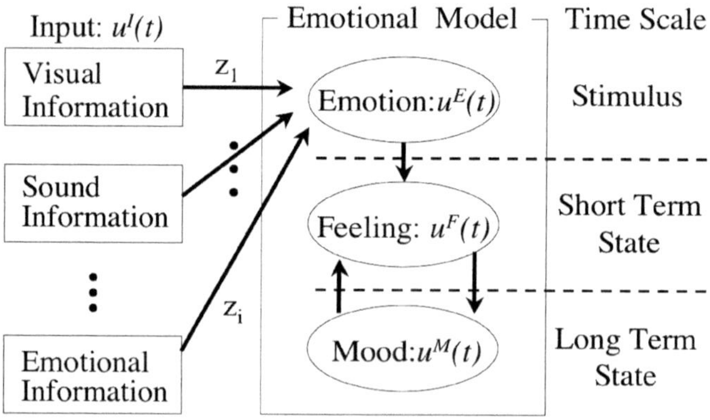

Fig. 4. The structure of the proposed emotional model

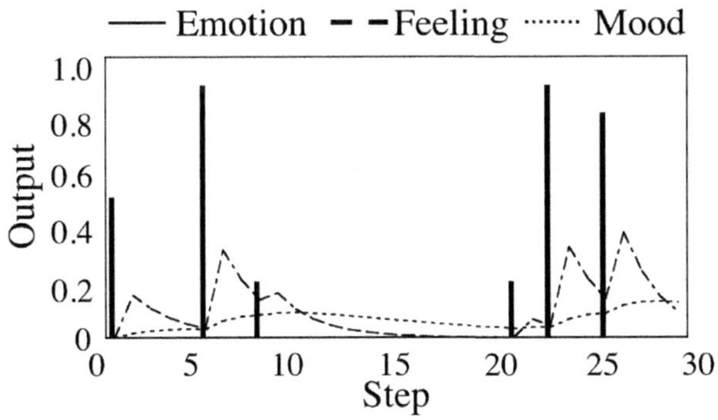

Fig. 5. An example of state change in emotional model

3.2 Emotion-Based Map Building

We use a temporal difference method based on state-value functions [34] for learning about the environment to implement location-dependent memory. The robot uses three state-value function maps of emotional memory, physical memory, and temporal reliability. Each state-value function is represented by a set of discrete cells [19].

In learning of the state-value functions based on location-dependent emotional memory, the robot determines the emotion value in a specific area. $v_i^F(x, y)$ corresponding to state of the ith feeling is updated as follows:

$$v_i^F(x_t, y_t) \leftarrow v_i^F(x_t, y_t) + \eta^F \left[r_{t+1}^F + \gamma^F v_i^F(x_{t+1}, y_{t+1}) - v_i^F(x_t, y_t) \right] \tag{7}$$

where η^F is a learning coefficient and γ^F a discount rate. A reward is given when a specific emotion occurs to the robot.

In the learning of the state-value function based on the location-dependent physical memory, the robot determines of the value place based on the experience of the past. The value of place $v_i^P(x, y)$, corresponding to the ith behavior mode, is updated as follows:

$$v_i^P(x_t, y_t) \leftarrow v_i^P(x_t, y_t) + \eta^P \left[r_{t+1}^P + \gamma^P v_i^P(x_{t+1}, y_{t+1}) - v_i^P(x_t, y_t) \right] \tag{8}$$

where η^P is a learning coefficient and γ^P is a discount rate. A reward is given when the robot detects objects such as users, balls, and battery-chargers. The robot uses a temporal reliability map with values and $v_i^R(x, y)$ to realize decision-making in a dynamic environment. The values of the sensed area are updated as:

$$v_i^R(x_s, y_s) = 1, \qquad (x_s, y_s) \in S \tag{9}$$

where S is the cell set corresponding to the sensed area. Values of remaining cells on the reliability map are updated by

$$v_i^R(x,y) \leftarrow \gamma^R v_i^R(x,y) \tag{10}$$

where γ^R is a discount rate.

3.3 Emotion-Based Multi-objective Behavior Coordination

The robot integrates visual tracking, collision avoidance, soccer playing, and wall following based on behavior mode and internal states. The basic behavior mode of robot is divided into random walking, battery-charging, user interaction, and soccer playing. Using a spiking neural network based on an internal state (see Section 3.1), the robot selects the behavior mode. Internal state $h_i(t)$ (eq.(4)) is expanded into

$$h_i(t) = h_i^{syn}(t) + h_i^{ext}(t) + h_i^{ref}(t) \tag{11}$$

$$h_i^{syn}(t) = \gamma^{syn} h_i(t-1) + \sum_{j=1, i \neq j}^{M} d_{j,i} o_j(t) \tag{12}$$

where γ^{syn} is the discount rate, $d_{j,i}$ is the weight connection strength from the jth to ith neuron, $o_j(t)$ is the spike output of the jth neuron at discrete time t, and M is the number of behavior modes (M=4). The robot uses information on internal states such as feelings and the amount of battery rest energy as external input $h_i^{ext}(t)$ to the spiking neural network.

After selecting the behavior mode, the robot selects the target point where the behavior can be taken. Next, the robot makes action outputs based on the multi-objective behavior coordination. A robot has a set of behaviors represented by fuzzy controller. A basic behavior of the robot can be represented using fuzzy rules based on simplified fuzzy inference [35]. In general, a fuzzy if-then rule is described as follows,

If p_1 is $A_{i,1}$ and ... and p_m is $A_{i,m}$
Then q_1 is $c_{i,1}$ and ... and q_n is $c_{i,n}$

where $A_{i,j}$ and $c_{i,k}$ are a Gaussian membership function for the jth input and a singleton for the kth output of the ith rule; m and n are the numbers of inputs and outputs, respectively. Fuzzy inference is described by,

$$\mu_{A_{i,j}}(p_j) = \exp\left(-\frac{(p_j - a_{i,j})^2}{b_{i,j}^2} \right) \tag{13}$$

$$\mu_i = \prod_{j=1}^{m} \mu_{A_{i,j}}(p_j) \tag{14}$$

$$q_k = \frac{\sum_{i=1}^{R} \mu_i c_{i,k}}{\sum_{i=1}^{R} \mu_j} \tag{15}$$

where $a_{i,j}$ and $b_{i,j}$ are the central value and the width of the membership function $A_{i,j}$; R is the number of rules. Outputs of the robot are motor output levels. Fuzzy controller is used for collision avoidance and target tracing behaviors. Here if the degree of the angry of the robot increases, the robot tries to avoid any interaction with people, balls, and others. Therefore, the value of interaction map; $V_i^P(x_t, y_t)$ is used as the input to fuzzy control in addition to the distance information to obstacles. And also, if the degree of the sadness of the robot increases, the robot tries to find people, balls, and others. Therefore, the value of interaction map; $V_i^P(x_t, y_t)$ is used as the input to fuzzy control in addition to the target direction. Furthermore, if the degree of the happiness of the robot increases, the robot tries to move toward the unknown areas. Therefore, the value of interaction map; $V_i^R(x_t, y_t)$ is used as the input to fuzzy control.

Next, we extend this fuzzy controller to multiple behaviors. We have proposed a multi-objective behavior coordination mechanism (Fig.6) [19]. A behavior weight is assigned to each behavior. The output is calculated by

$$q_k = \frac{\sum_{h=1}^{B}\left(w_h(t) q_{h,k}\right)}{\sum_{h=1}^{B} w_h(t)} \tag{16}$$

where B and $w_h(t)$ are the number of behaviors and a behavior weight of the hth behavior over the discrete time step t, respectively. By updating the behavior weights based on the perceptual information, the robot can take a multi-objective behavior according to the time series of perceptual information. This method can be considered as a mixture of experts, because the behavior coordination mechanism is considered as a gating network.

In this chapter, we use collision avoidance and target tracing behaviors. The behavior weights are updated according to the perceptual information in our previous research [19], but they are updated according to the location-dependent memory and feelings of the robot. Therefore, the perceptual information is indirectly used for the behavior coordination in this chapter. The behavior weight of collision avoidance is calculated by

$$w_1(t) = \frac{1}{2}\left(v_i^F(x_t, y_t) + u_i^F(t)\right) \tag{17}$$

where $v_i^E(x_t, y_t)$ is the value of the degree of danger at the current position in the map; the ith feeling is corresponding to the fear. In order to simplify the problem, we use the following behavior weight for target tracing,

$$w_2(t) = 1 - w_1(t) \tag{18}$$

In this way, the robot can take actions according to the state of feelings based on location-dependent memory.

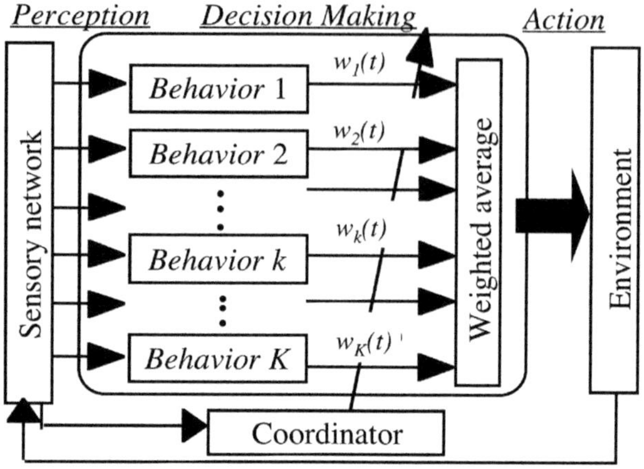

Fig. 6. Multi-objective behavior coordination

4 Experimental Results

To confirm the feasibility of our proposed method, we conducted experiments using the emotional model of a partner robot in VRE. The VRE in this example includes a table, four chairs, a battery-charger, two balls, and two users. Perceptual information available to the robot includes distance information measured in 8 directions by ultrasonic sensors, map information, the locations of users and ball in the sensing range of the VRE in addition to internal emotional model states. The robot interacts with a user in a VRE using a CCD camera and microphone (see Section 2.1). If the robot recognizes words via speech recognition through the PC microphone, for example, it attempts to detect a user in the VRE. In this experiment, the number of utterances in the database is 100, and the number of words in the database used for voice recognition is 150. The robot moves in a simulator corresponding to the VRE (Fig. 7). When encountering a battery, a user, or a ball, the robot acts based on the internal model. Based on the user reaction and perceptual environment information, the emotional model updates the states of happiness, anger, sadness, and fear as feeling parameters, and updates the amount of rest energy and the environmental map based on perceptual information.

We conducted experiments comparing the trajectories of the robot without the emotional model (Case 1) and with it (Case 2) (Fig. 8). The robot tried to avoid the dangerous area by referring to the environmental map based on location-dependent emotional memory. This experimental result shows the effectiveness of the emotional model in order to evaluate the temporal and positioning information. Next, we conducted experiments on the reliability of the environmental information in the decision making (Figs.9-1 and 9-2). The depth of gray indicates the temporal reliability in the map. In this result, the robot tried to select the reliable target location

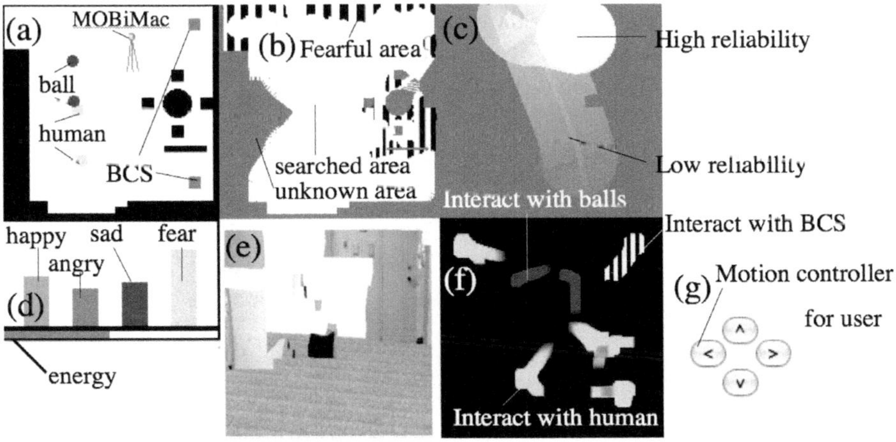

Fig. 7. Interaction in VRE; (a) plain map, (b) map of location-dependent emotional memory, (c) map of temporal reliability, (d) emotional model, (e) 3Dmap, (f) map on physical interaction with users (g) motion controller for a user in the VRE

(T_2) from two candidate positions of the charging stations by referring to the temporal reliability map. Next, we show an experimental result between the robot and a person in VRE (Fig.10). We used the voice recognition through PC in VRE. In Fig.10 (a), the person A asked the robot to come here, but the robot tried to pass by the person B after the degree of happiness increases. At the time, the person A told the robot "dangerous". Communication such as verbal interaction and body language evolved to inform others of danger, e.g., the word "dangerous" or the sentence "go away." As a result, the robot updated the area around the person B and ball as the fear area according to the feeling parameters. The experimental results show that the emotion incentive function enables to take the actions requested by the sender's emotion. The emotional model can realize the actions through the interaction dependent on the situation, and the sender in the RE can perform human-like interaction, because the human can easily predict the reaction of the robot using the emotional model similar with a human.

Finally, we conducted experiments on the behavior selection. Figure 13-1,2 shows the situation before the robot selects a target. The battery level of the robot is almost full because the robot is on the battery. As a result, the degree of happiness becomes high (Fig.13-1 (c)). At the time, the reliability of upper-right region is very low. Therefore, the robot selects the target point in the upper-right region (Fig.13-2 (b)). The selected target point is drawn by an yellow circle. Figure 13-2(c) shows the trajectory of the robot. The robot tried to search in the unknown area with considering a risk of battery exhausting according to the state of emotions. The other experimental results related with the proposed method are shown in [19]. In this way, the emotional model is useful and effective to realize the multi-objective decision making in order to evaluate the situation.

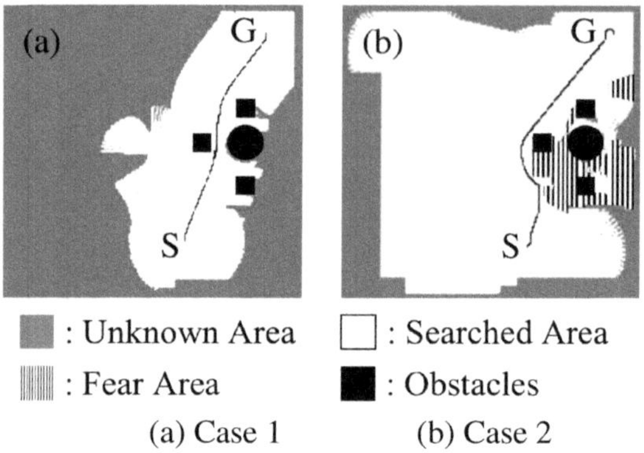

Fig. 8. Collision avoidance based on the emotional map; S is the starting point and G is the target point

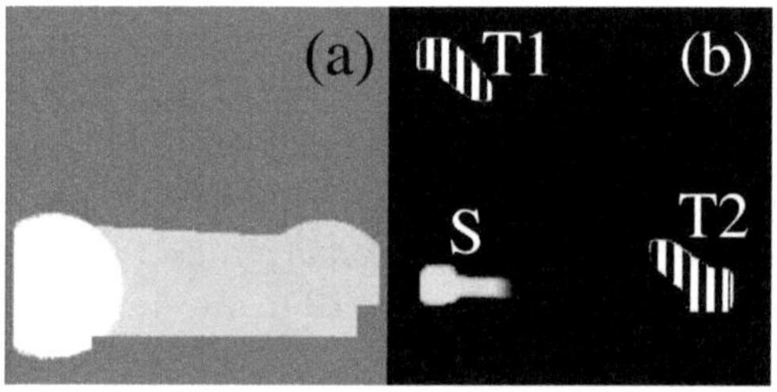

Fig. 9-1. Robot trajectory based on the map of temporal reliability; S is the starting point. T_1 and T_2 is two candidates of target point.

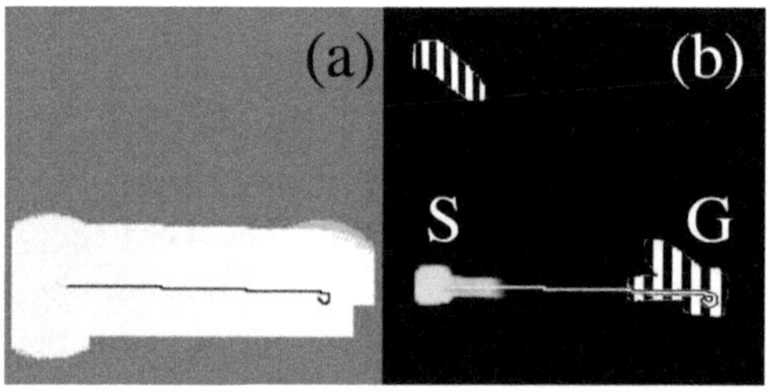

Fig. 9-2. Robot trajectory based on the map of temporal reliability; S is the starting point and G is the selected target point from two candidates

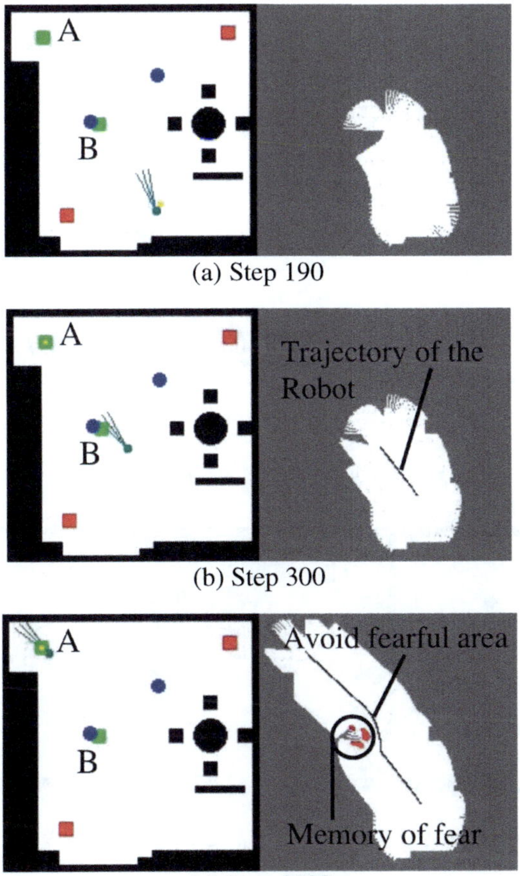

(a) Step 190

(b) Step 300

(c) Step 1100

Fig. 10. Example of robot/user interaction

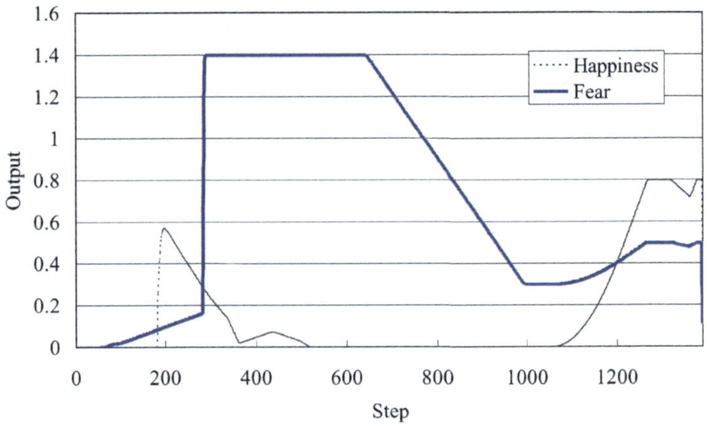

Fig. 11. History of change in feelings

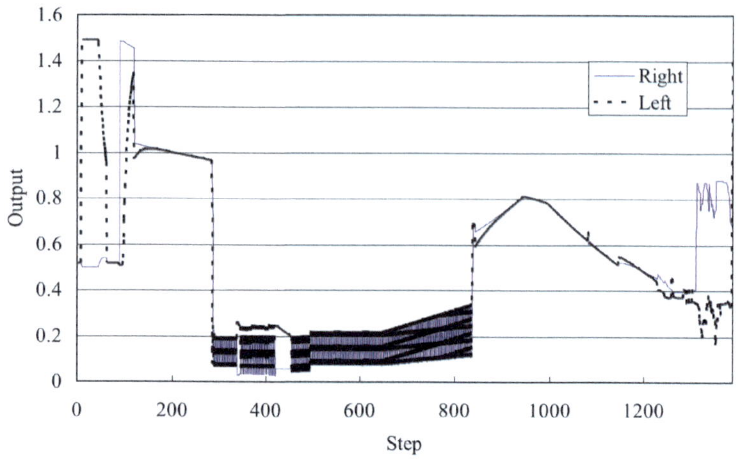

Fig. 12. History of change in motor output

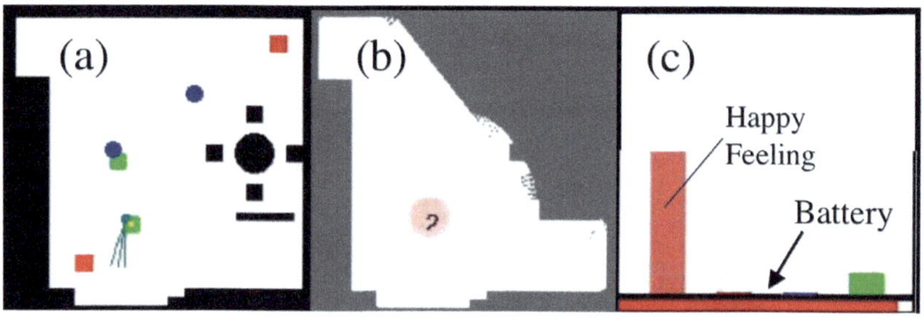

Fig. 13-1. The trajectory of the robot based on happy feeling

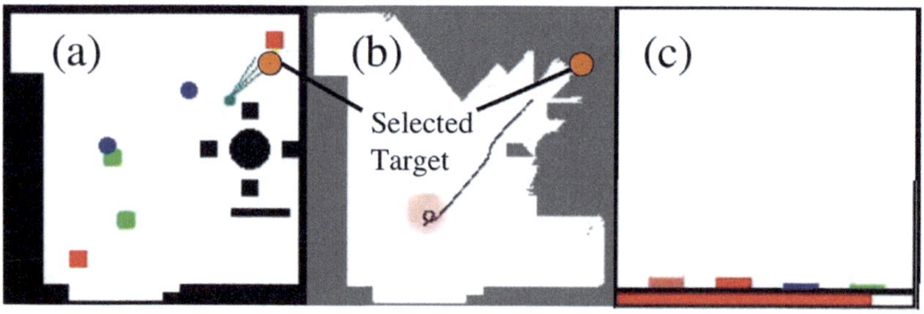

Fig. 13-2. The trajectory of the robot based on happy feeling

5 Summary

In this chapter, we discussed the availability of emotional model for partner robots based on computational intelligence, and furthermore, the effectiveness of the emotional model for a partner robot based on location-dependent memory. The computational intelligence technologies can be considered as strategies in the decision making and learning, while the emotional model can be used for evaluating the state of decision making and learning in the tactics. In this way, we can synthetically integrate these two roles in the robot. First, we applied state-value functions to represent location-dependent memory of feelings, physical interaction, and reliability. Furthermore, we proposed a multi-objective behavior coordination method based on the emotional model.

The experimental results of collision avoidance, trajectory, target selection, and communication demonstrated the effectiveness of our proposed method in realizing experience-based, location-dependent memory. The main advantage of our proposed method is in using an emotional model similar to a user, not in the preciseness of the emotional model. Feeling parameters are very easy for the others to understand the meaning of the evaluation factors. Therefore, we can design collision avoidance, target selection, and communication with users easily. Furthermore, feeling parameters are easy to use in unified indexes for representing internal states of robot. To share situations with users, robots should understand the meaning of event sharing, *i.e.*, it should be able to realize informative, evocative, and incentive functions of emotion. We have not yet discussed how to share the meaning of events and experiences. Therefore, we intend to develop the understanding of events surrounding the robot based on image processing, speech recognition, and sensor

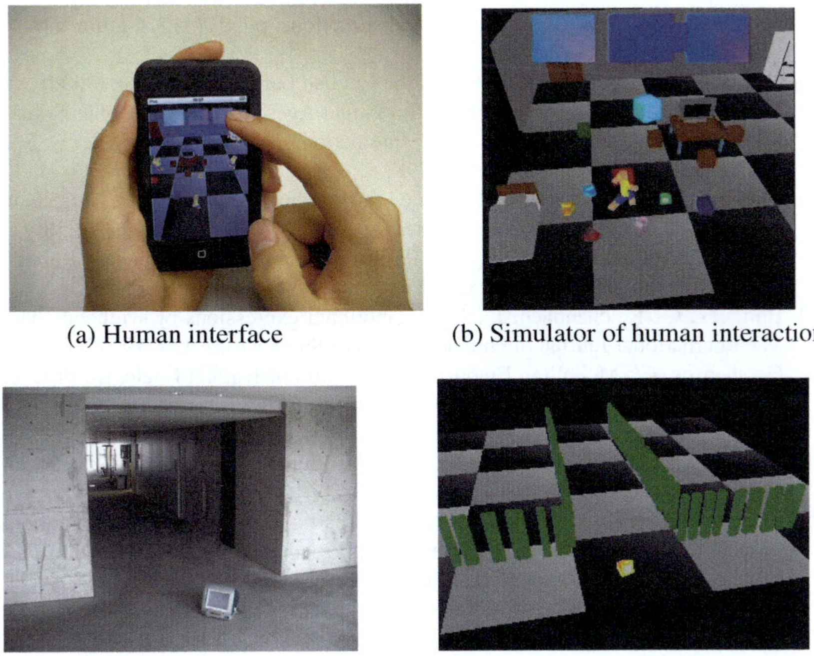

(a) Human interface (b) Simulator of human interaction

(c) Real Environment (d) Environmental map based on SLAM

Fig. 14. Human interaction based on iPhone and iPod touch measurement. Furthermore, we are developing a human interface based on iPhone and iPod touch (Fig.14) [38]. The environmental map is generated by the simultaneous localization and mapping by the robot [36]. We discuss the emotional interaction with people through the devices as a future work.

References

[1] Sperber, D., Wilson, D.: Relevance - Communication and Cognition. Blackwell Publishing Ltd., Malden (1995)

[2] Cornelius, R.R.: The Science of Emotion: Research and Tradition in the Psychology of Emotion, pp. 149–184. Prentice Hall College Div, Englewood Cliffs (1995)

[3] Buck, R.: The Communication of Emotion. Guilford Press, New York (1988)

[4] Keltner, D., Haidt, J.: Social functions of emotions. In: Mayne, J., Bonann, G.A. (eds.) Emotion and Adaptation. Oxford University Press, Oxford (2001)

[5] Keltner, D., Kring, A.M.: Emotion, Social Function, and Psychopathology. Review of General Psychology Copyright 1998 by the Educational Publishing Foundation 2(3), 320–342 (1998)

[6] Ridley, M.: The Origins of Virtue. Felicity Bryan, Oxford (1996)

[7] Eysenck, M.W.: Psychology: An Integrated Approach, Harlow, Essex, Longman (1998)

[8] Cosmides, L., Tooby, J.: Evolutionary psychology and the emotions. In: Lewis, M., Haviland-Jones, J.M. (eds.) Handbook of Emotions, pp. 91–115. Guilford, New York (2000)

[9] The American Heritage Dictionary of the English Language, 4th edn. (2000)

[10] El-Nasr, M.S., Skubic, M.: A Fuzzy Emotional Agent for Decision-Making in A Mobile Robot. In: Proc. of the 1998 International Conference on Fuzzy Systems, An-chorage, Alaska, pp. 135–140 (1998)

[11] Wada, K., Shibata, T., Saito, T., Tanie, K.: Psychological and Social Effects of One Year Robot Assisted Activity on Elderly People at a Health Service Facility for the Aged. In: Proc. (CD-ROM) of 2005 IEEE International Conference on Robotics and Automation, pp. 2796–2801 (2005)

[12] Bartneck, C., Reichenbach, J.: Subtle emotional expressions of synthetic characters. The International Journal of Human-Computer Studies (62), 179–192 (2005)

[13] Gershenson, C.: Modelling Emotions with Multidimensional Logic. In: Proc. of 18th International Conference of the North American Fuzzy Information Processing Society, New York City, pp. 42–46 (1999)

[14] Russell, J.A.: A circumplex model of affect. Journal of Personality and Social Psychology 39, 1161–1178 (1980)

[15] Kubota, N., Omote, S., Mori, Y.: Emotional Learning of A Vision-Based Partner Robot for Natural Communication with Human. In: Proc. (DVD-ROM) of 2006 IEEE World Congress on Computational Intelligence (WCCI 2006), Vancouver, Canada, July 16-21, 2006, pp. 6288–6294 (2006)

[16] Kubota, N., Omote, S., Mori, Y.: Emotional Model of A Vision-Based Partner Robot for Natural Communication with Human. In: Proc (CD-ROM) of The 3rd International Conference on Computational Intelligence, Robotics and Autonomous Systems, INV104-104002 (2005)

[17] Kubota, N., Pei, J., Mori, Y.: An Emotional Model for Natural Communication of A Partner Robot. In: Proc. of the 8th International Symposium on advanced Intelligent System, vol. TA03-1, pp. 153–159 (2007)
[18] Kubota, N., Nojima, Y., Baba, N., Kojima, F., Fukuda, T.: Evolving Pet Robot with Emotional Model. In: Proc. of the 2000 Congress on Evolutionary Computation, La Jolla, CA, pp. 1231–1237 (2000)
[19] Kubota, N., Wakisaka, S.: An Emotional Model Based on Location-Dependent Memory for Partner Robots. Journal of Robotics and Mechatronics 21(3), 317–323 (2009)
[20] Kubota, N., Shimizu, T., Abe, M.: Joint Attention of A Partner Robot Based on Computational Intelligence. In: Proc. (CD-ROM) of 2nd International Symposium on Computational Intelligence and Intelligent Informatics (ISCIII 2005), pp. 196–201 (2005)
[21] Kubota, N., Urushizaki, Y.: Communication Interface for Human-Robot Partnership. Journal of Robotics and Mechatronics 16(5), 526–534 (2004)
[22] Kubota, N., Satomi, M., Taniguchi, K., Nogawa, Y.: Human Three-Dimensional Modeling based on Intelligent Sensor Fusion for A Tele-operated Mobile Robot. In: Proc. (CD-ROM) of 11th International Conference, Knowledge-Based Intelligent Information and Engineering Systems, pp. 98–113. Vietri sul Mare, Italy (2007)
[23] Kubota, N., Nishida, K.: Fuzzy Computing for Communication of A Partner Robot Based on Imitation. In: Proc. (CD-ROM) of 2005 IEEE International Conference on Ro-botics and Automation, pp. 4391–4396 (2005)
[24] Kubota, N., Nishida, K.: Human Recognition of A Partner Robot Based on Relevance Theory and Neuro-Fuzzy Computing. In: Proc. (CD-ROM) of 2005 IEEE/RSJ Interna-tional Conference on Intelligent Robots and Systems, Edmonton, Alberta, Canada, August 2-6, 2005, pp. 2812–2817 (2005)
[25] Kubota, N., Nishida, K.: Development of Internal Models for Communication of A Partner Robot Based on Computational Intelligence. In: Proc. of 6th International Symposium on Advanced Intelligent Systems, pp. 577–582 (2005)
[26] Bezdek, J.C.: What is Computational Intelligence? In: Zurada, J.M., Marks II, R.J., Robinson, C.J. (eds.) Computational Intelligence - Imitating Life, pp. 1–12. IEEE Press, Los Alamitos (1994)
[27] Fukuda, T., Kubota, N.: Computational Intelligence for Robotic Systems. In: Kaynak, O., Tosunogle, S., Ang Jr., M. (eds.) Recent Advances in Mechatronics, pp. 13–26. Springer, Heidelberg (1999)
[28] Ballard, D.H.: An Introduction to Natural Computation. The MIT Press, Cambridge (1997)
[29] Lee, C.C.: Fuzzy Logic in Control Systems: Fuzzy Logic Controller - Part I & Part II. IEEE Trans. on Systems, Man, and Cybernetics 20(2), 404–435 (1990)
[30] Fogel, D.B.: Evolutionary Computation. IEEE Press, Los Alamitos (1995)
[31] Kubota, N., Arakawa, T., Fukuda, T., Shimojima, K.: Motion Planning for A Robotic System with Structured Intelligence. In: Proc. of the 1997 IEEE International Symposium on Computational Intelligence in Robotics and Automation, pp. 60–65 (1997)
[32] Sutton, R.S., Barto, A.G.: Reinforcement Learning. The MIT Press, Cambridge (1998)
[33] Gerstner, W.: Spiking Neurons. In: Maass, W., Bishop, C.M. (eds.) Pulsed Neural Networks, ch. 1, pp. 3–53. MIT Press, Cambridge (1999)
[34] Sutton, R.S.: Learning to Predict by the Methods of Temporal Differences. Machine Learning 3, 9–44 (1988)

[35] Jang, J.S., Sun, C.T., Mizutani, E.: Neuro-Fuzzy and Soft Computing. Prentice-Hall, Inc., Englewood Cliffs (1997)
[36] Sasaki, H., Kubota, N., Taniguchi, K., Nogawa, Y.: Illuminance Measurement and SLAM of a Mobile Robot Based on Computational Intelligence. In: Proc. (CD-ROM) of 7th IEEE International Symposium on Computational Intelligence in Robotics and Automation (CIRA 2007), June 20-23, 2007, pp. 541–545 (2007)
[37] Hatano, G., Okada, N., Tanabe, H. (eds.): Affective Minds, pp. 101–104. Elsevier Science, B. V. (2000)
[38] Kubota, N., Wakisaka, S., Yorita, A.: Tele-operation of Robot Partners through iPod touche. In: 4th International Symposium on Computational Intelligence and Intelligent Informatics (ISCII 2009), October 21-25, pp. 75–80 (2009)

Chapter 6
Facial Expression Synthesis Based on Emotion Dimensions for Affective Talking Avatar

Shen Zhang[1], Zhiyong Wu[1,2], Helen M. Meng[3], and Lianhong Cai[1]

[1] Key Laboratory of Pervasive Computing, Ministry of Education
Tsinghua National Laboratory for Information Science and Technology
Department of Computer Science and Technology,
Tsinghua University, Beijing 100084, China
[2] Tsinghua-CUHK Joint Research Center for Media Sciences,
Technologies and Systems,
Graduate School at Shenzhen, Tsinghua University, Shenzhen
[3] Department of Systems Engineering and Engineering Management
The Chinese University of Hong Kong, HKSAR, China
zhangshen05@mails.tsinghua.edu.cn, john.zy.wu@gmail.com,
hmmeng@se.cuhk.edu.hk, clh-dcs@tsinghua.edu.cn

Abstract. Facial expression is one of the most expressive ways for human beings to deliver their emotion, intention, and other nonverbal messages in face to face communications. In this chapter, a layered parametric framework is proposed to synthesize the emotional facial expressions for an MPEG4 compliant talking avatar based on the three dimensional PAD model, including pleasure-displeasure, arousal-nonarousal and dominance-submissiveness. The PAD dimensions are used to capture the high-level emotional state of talking avatar with specific facial expression. A set of partial expression parameter (PEP) is designed to depict the expressive facial motion patterns in local face areas, and reduce the complexity of directly manipulation of low-level MPEG4 facial animation parameters (FAP). The relationship among the emotion (PAD), expression (PEP) and animation (FAP) parameter is analyzed on a virtual facial expression database. Two levels of parameter mapping are implemented, namely the emotion-expression mapping from PAD to PEP, and the linear interpolation from PEP to FAP. The synthetic emotional facial expression is combined with the talking avatar speech animation in a text to audio visual speech system. Perceptual evaluation shows that our approach can generate appropriate facial expressions for subtle and complex emotions defined by PAD and thus enhance the emotional expressivity of talking avatar.

Keywords: facial expression, emotion, talking avatar, PAD.

1 Introduction

In face-to-face interactions, people expressed themselves through a number of different modalities, such as speech, facial expression, body gestures etc. Speaking is not only moving the lips, but also includes the eye-contact, eyebrow-raise, and

T. Nishida et al. (Eds.): Modeling Machine Emotions for Realizing Intelligence, SIST 1, pp. 109–132.
springerlink.com

other facial movements. Facial expression is one of the most effective ways for human to express their intentions, attitude, inner emotional states and other nonverbal messages in speech communications [13]. Early efforts in psychological research have established that the affective information in human communications is delivered at different ratio by different modalities: the verbal expression (e.g., words, spoken text) only accounts for 7% of the affective meaning of speakers' feeling and attitude. The vocal expression (e.g., prosody, stress) conveys 38% of the affective message. By marked contrast, the facial expression accounts for 55% of the affective information [32]. Especially when inconsistent messages are expressed by different modality, the facial expression tends to be more reliable for human perception of emotion [35]. Facial expression is also considered as the one of the most important aspect in analyzing and modeling natural human emotions [40]. The study of facial expression has attracted broad attentions from psychologists [39, 45], biologists [10], computer engineers [53], and artists [15]. The reason why facial expression receives so many interests is not only because it plays important roles in human communications, but also its complexity and difficulty to analyze and synthesize.

Talking avatar, an animated speaking virtual character with vivid human-like appearance and natural synthetic speech, has gradually shown their potential application in human computer intelligent interactions, because of its communication abilities to deliver rich verbal and nonverbal information by voice, tones, eye-contact, body gestures and facial expressions, etc. Possible applications of talking avatar include a virtual storyteller for children [21, 51], a virtual guider or presenter for personal or commercial website [36, 56], a representative of user in computer game [29] and a funny puppetry for computer mediated human communications [42]. It is clearly promising that the talking avatar will become an expressive multimodal interface in human computer interaction. However, current research on human emotion has not yet advanced to the stage where it is possible to synthesize an affective intelligent talking avatar who can express their feelings and emotions through vocal and facial behaviors as natural as human beings. To this end, in our research on expressive talking avatar synthesis, it is necessary to generate appropriate facial expression according to the emotional content in speech in order to enhance the affective expressivity of talking avatar. Our long-term goal is to implement an expressive talking avatar with lip-articulation and emotional facial expressions for text-to-audio-visual-speech (TTAVS) system.

In this chapter, a parametric layered framework is proposed to synthesize emotional facial expression for an MPEG-4 compliant talking avatar based on psychological emotion dimensions. The following three research questions will be addressed in our work:

1) How to describe various emotional states in a quantitative and unified way?
2) How to model the facial expression based on animation control parameters?
3) What is the interrelation between emotional state and facial expression?

Quick answers to these questions are that we adopt the PAD psychological model to describe arbitrary emotional state by three dimensions, namely pleasure-displeasure (P), arousal-nonarousal (A) and dominance-submissiveness (D) [33].

A set of Partial Expression Parameters (PEP) is proposed to depict the typical facial expressive movement in face regions based on the correlation among MPEG-4 facial animation parameters (FAP). The PAD and PEP mapping function and the PEP-FAP translation function are trained on a relatively small facial expression database. A parametric layered framework is thus established with the PAD as high-level emotion descriptor, the PEP as mid-level facial expression configuration and the FAP as low-level controller of face model deformation on three-dimensional avatar.

The chapter is organized as follows. Section 2 presents the literature review on the emotion description and facial expression synthesis, and introduces the PAD emotional model. The layered framework for parametric facial expression synthesis based on emotional dimensions (i.e. pleasure-arousal-dominance) is described in Section 3. Details of PEP definition and its translation to FAP are illustrated in Section 4. In Section 5, the establishment of pseudo facial expression database on talking avatar is reported. The interrelation between emotional state description (PAD) and facial expression configuration (PEP) of talking avatar is investigated in Section 6. The integration of the proposed facial expression module into text-to-audio-visual-speech system is described in Section 7. The resulting emotional facial expression synthesized by the proposed PAD-PEP-FAP framework and a series of perceptual evaluation are illustrated in Section 8. Finally, we conclude our work and discuss the future direction.

2 Related Work

Facial expression and its relationship with emotion are topics having been actively studied. Early academic study on facial expression may date back to the contribution made by Darwin [10]. Since then emotion and facial expression are studied mainly in psychological area. The development of facial action coding system (FACS) has accelerated the research of automatic facial expression analysis in the filed of computer vision since 1978 [27]. After Picard proposed the theory of affective computing in 1997 [40], facial expression is regarded as one of the most important visual channels to perceive and express human emotion. It is necessary for affective computer to understand and perform appropriate facial expression in human computer communication. From the aspect of talking avatar synthesis, in this section, we review the previous work on emotion description by psychologists, and facial expression synthesis by computer scientists. In addition, for parametric facial animation, we introduce two important coding systems, namely the FACS proposed by Ekman [14] and facial animation framework in MPEG-4 standards [34].

2.1 Categorical Emotion Description

How to describe human emotion and facial expression appropriately and effectively? This question has been discussed by scientists (mainly psychologists) for a long time. Various existing emotion descriptive models can be summarized into two aspects: the categorical and dimensional.

In his research on human expression of emotions, Darwin worked with a set of "state of mind", which includes emotion (e.g. anger, jealousy, terror, love), motivational, behavioral or personality traits (e.g. determination, defiance, ambition), sensations (e.g. bodily pain, hunger) and cognitive processes (e.g. abstraction, meditation). In addition, Darwin proposed that the expressions of "high spirits" and "low spirits" are best recognized [45]. The "state of mind" and "high/low spirits" can be regarded as a mixture of categorical and dimensional approach to describe emotion.

The "basic emotion" theory proposed by Ekman et al. has great influences on psychological research on facial expression. This theory is centered on a small set of discrete categories, which includes happiness, sadness, surprise, fear, anger, and disgust, namely the "big six" emotions. These basic emotions are considered to be universally perceived with respect to facial expression, regardless of culture [12]. The "big six" emotions is perhaps the most popular categorical approach to describe emotional state. The advantage of categorical approach lies in that it is the most straightforward and simplest way to describe facial expression in daily life. The existence of large amount of emotion-denoting words in natural languages makes the categorical approach powerful, intuitive and consistent with human's experiences [57]. However, the vagueness and ambiguity of natural language make it hard to clearly recognize the subtle variation of spontaneous occurring emotions. It is also difficult to describe the continuum or non-extreme emotional state by discrete list of categories. Although the basic emotion categories are widely adopted in facial expression analysis and recognition [27], when it comes to synthesis, the basic emotions only cover a relatively small set of prototypical facial expressions. For expressive talking avatar who aims to speak, move and behave as naturally as human, the expression of basic emotions or discrete emotion categories are not enough, while a unified and flexible approach is needed to model the complex and subtle facial expressions in human life.

2.2 Dimensional Emotion Description

An alternative approach is the dimensional description of emotion. For practical and theoretical purpose, early efforts attempted to describe emotion based on a hybrid of both basic-to-complex categories and emotional primitives. The Plutchik's wheel (2D) and "cone-shaped model" (3D) are proposed to describe emotions by exploring their relations in terms of similarity and intensity [43]. The 3D emotion cone's vertical dimension represents intensity, and the circle represents degrees of similarity among different emotions. There are eight primary emotions and other emotions occur as combinations, mixtures, or components of the primary emotions. The difference among emotions is measured by their degree of similarity to one another. Emotions can vary in degrees of intensity or levels of arousal. The Plutchik's theory tried to uncover the relations among different emotions by dimensional analysis, but is still limited in categorical concept and thus not a completely dimensional model.

For dimensional emotion description, ideally, arbitrary emotional state can be quantitatively measured in terms of a small number of basic dimensions in a multi-dimensional space. Many researchers try to propose different emotion dimensions to describe the essential property of various emotional state, and three basic dimensions, with different names, can be summarized from previous research [7, 37, 44]. In this study, we choose the following terms pleasure (i.e. valence, evaluation), arousal (i.e. activation, activity), dominance (i.e. power, potency). The pleasure dimension distinguishes the positive-negative quality of emotional state, the arousal dimension measures the degree of physical activity and mental alertness of emotional state, and the dominance dimension is defined in terms of control versus lack of control, mainly focus on the social aspect of emotion. Based on these basic dimensions, different emotional models are proposed and explored in multi-disciplinary area. Cowie et al. proposed 2D disk-shaped activation-evaluation emotion space [8]. Mehrabin et.al developed the PAD emotional model by which all emotions can be located in 3D emotion space where the pleasure, arousal and dominance dimensions are orthogonal and independent to each other, as shown in Figure 1 [28, 33].

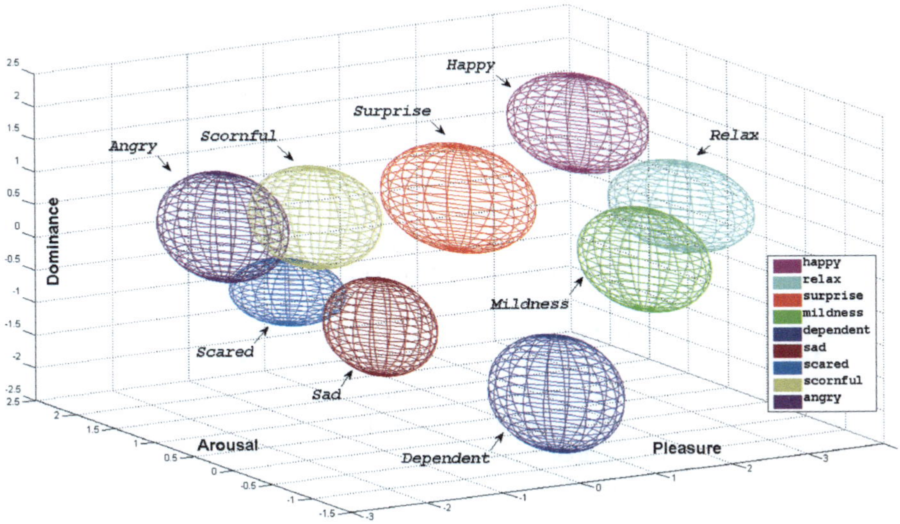

Fig. 1. The distribution of 9 emotional states/words in PAD emotion space. The center of each ellipsoid is the mean and the radius is standard deviation of each emotional state. The PAD data is provided by Institute of Psychology, Chinese Academy of Sciences [28].

The emotion dimensions can measure the continuous variation of emotional states, but they are not so intuitive as emotion labels in natural language. To this end, different rating methods and tools are developed to annotate the value of emotional dimension, such as the self assessment manikins (SAM) tool [25], the Feeltrace system [8], and the abbreviated PAD emotion scales [28]. The use of dimensional emotion description has been explored in multimodal affect recognition and synthesis [3, 61], emotional speech synthesis [48], facial expression recognition [7]. However, research on facial expression synthesis based on emotion dimensions is rare. In this chapter, we explore the use of PAD emotional model to synthesize continuum of varied facial expression for talking avatar. According to Mehrabian's theory [33], human emotion is not limited to isolated categories but can be described along three nearly orthogonal dimensions: pleasure-displeasure (P), arousal-nonarousal (A) and dominance-submissiveness (D), namely the PAD emotional model which is proven to be appropriate in describing emotions. The PAD emotional model has been successfully applied in acoustic correlates analysis of emotional speech [9] and expressive speech synthesis [59, 60].

2.3 Emotional Facial Expression Synthesis

Previous research on facial expression synthesis adopted the basic emotion theory [11, 30, 58], namely the "big six". The reason why basic emotion is widely used is not only because it is simple and intuitive, but also because most of the data used in early research are series of deliberate facial expression with extreme intensity rather than spontaneous occurring facial expression in natural conditions. For facial expression synthesis, most research focus on realistic-looking facial animation to enhance the naturalness while neglect delivering the inner emotional state or communication message by emotional facial expression. Ruttkay et al.[46] proposed the "Emotion Disc" and "Emotion Square" to generate continuum of emotional facial expression by two-dimensional navigation. Albrecht et al. [2] extended the work of MPEG-4 based facial expression synthesis [53], and designed a method to generate mixed expressions where the two-dimensional "Activation-Evaluation" emotion space is explored. Kshirsagar et al. [24] have proposed a multi-layered framework for facial expression synthesis which consists of high-level description of dynamic emotion intensity, mid-level configuration of static facial expressions, and low-level FAPs definition. Exploration on the paralinguistic meaning of facial movements is undertaken for isolated facial area, such as eyebrow and eye-lid [19]. However, there are few research results on the correlation between emotional dimensions and facial movements for synthesis purpose, at least not as much as the study of correlation between emotional dimensions and acoustic features for emotional speech synthesis [47]. To this end, the quantitative study on facial correlates with emotional dimensions is conducted in this chapter to realize the parametric facial expression synthesis for affective talking avatar.

2.4 FACS and FAP

The description and measurement of facial feature movement is an essential aspect to facial expression synthesis. The emotional sate is usually expressed by subtle change of facial movement in local area, while the prototypical facial expression with extreme intensity occurs relatively infrequently. To capture the local and subtle facial feature movements, Ekman et al. proposed facial action coding system (FACS), which is a human-observer-based system designed to capture the subtle movement of isolated facial features. The basic facial motion pattern is called "action unit" (AU) in FACS[14]. There are 44 AUs defined in FACS, among which 30 AUs are anatomically related to contraction of a specific set of facial muscles, and the other 14 AUs are referred as miscellaneous actions. Trained experts can use these AUs individually or by their combinations to describe facial expression with 5-point ordinal scale measuring the intensity of muscle contraction. Another set of emotion-specified facial action coding system are proposed as EMFACS [17]. The FACS is originally designed to be human-readable not machine-readable, therefore there are many attempts to translate facial features movements to AUs for automatic facial expression analysis [6, 16, 52]. From the perspective of computer animation, the action units define the motion patterns of facial features, but not provide the quantitative temporal and spatial information required by face animation [20]. Therefore, the FACS system is widely adopted in filed of facial expression recognition, but it is not suitable for synthesis and animation purpose.

On the contrary to FACS, the facial animation framework developed under MPEG4 standards is designed completely from the perceptive of computer animation [34]. In accordance to action units, the facial animation parameters (FAPs) are designed based on the study of minimal perceptible actions and are closely related to sets of muscle actions [22, 38, 50, 55]. 84 facial feature points are specified to provide spatial references for defining 68 FAPs as quantitative movement of specific facial feature points. The FAPs represent a complete set of basic facial actions including head motion, tongue, eye, and mouth control, and thus it can be used to control synthetic face model to perform various kinds of facial expressions with varied intensity, even the exaggerated or distorted facial expression that only occurs on cartoon face.

From the view of facial expression synthesis, FAPs are much flexible and suitable to animate face models displaying various facial movements. However it is a tedious work to manipulate FAPs directly for generating emotional facial expression, since the FAPs control the facial feature points movement for low-level mesh deformation, rather than meaningful facial motion patterns, not mention the emotional facial expression. To this end, in this chapter, we propose the partial expression parameter (PEP) to depict the expressive movement of facial features (e.g. eyebrow, eye, mouth etc.), which combines the advantage of both FACS in human perceptible description of facial movement and FAP in low-level deformation of face model.

3 Layered Framework for Facial Expression Synthesis

To synthesize emotional facial expression for affective talking avatar, we designed a parametric layered framework based on emotion dimensions as Figure 2 shows. In this layered framework, the input is the high-level emotional parameters indicating the intended emotional state of the talking avatar, namely the pleasure-arousal-dominance (PAD) triple. The output is the expressive facial movement of virtual avatar. For geometric facial deformation, we adopt the facial animation parameters (FAPs), which is designed to control the face model points to generate various facial movements. However, it is too complicated for facial expression synthesizer to manipulate FAPs directly since they only define the motions (e.g. translations and rotations) of single facial feature points. To this end, a set of

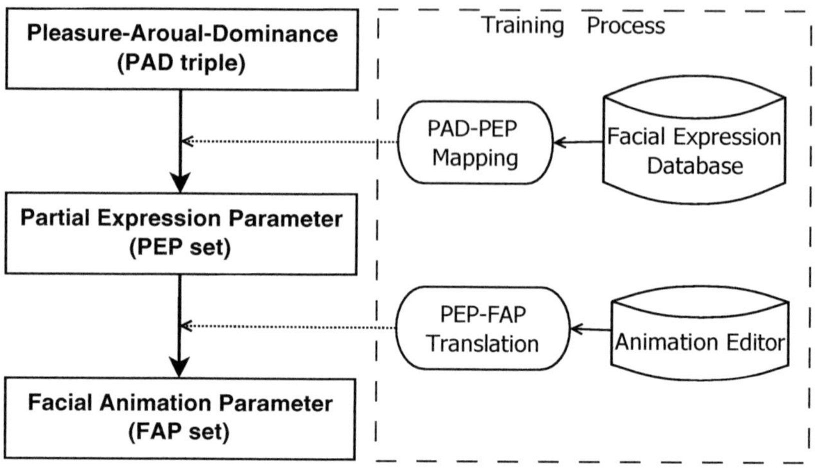

Fig. 2. Layered framework for PAD-driven facial expression synthesis

partial expression parameters (PEPs) are proposed to depict local emotional facial expressions, such as eyebrow-raise, mouth-bent and eye-open etc. The PEPs model the correlation among FAPs within local face region, and thus reduces the complexity of FAP-driven facial expression synthesis. To synthesize emotional facial expression, the PAD triple is first mapped to PEPs, after that the PEPs are translated to FAPs, finally the FAPs are used to animate the talking avatar to perform the facial expression which is expected to match the emotional state described by the input PAD triple.

With PAD as high-level description of emotional state, PEP as mid-level configuration of facial expression and FAP as low-level animation controller of face model, a layered framework for facial expression synthesis is implemented, where the PAD-PEP mapping model is trained on a facial expression database on talking avatar, and the PEP-FAP translation model is defined experimentally using a home-grown expression editor based on the study of FAP correlations [26] as well as the FACS manual [14]. In the following sections, we will introduce the approaches to implement the above layered framework and its integration to our emotional text to audio visual speech system [58].

4 PEP-FAP Translation Model

The partial expression parameter (PEP) is inspired by the study of FAP correlation and the observation of emotional facial expression within local face region. The PEP is proposed to depict the common emotional facial movement within specific facial regions. For low-level facial expression animation, a translation between PEP and FAP is needed. In this section, we describe the implementation of a linear interpolation to translate the PEP to its corresponding FAP subset.

4.1 Partial Expression Parameter (PEP)

As stated in section 2.4, the FAP is not very suitable as the configuration of emotional facial expression, since they only define the motions (e.g. translations and rotations) of single facial feature points. On the other side, there exist high correlations among different FAPs within the same facial regions as reported in the study of FAP interpolation algorithm for transmitting FAP in low bandwidth [26]. This motivates us to propose a new parameter to capture the FAP correlations as well as depict the common emotional facial movement in local facial regions, such as mouth-bent, eye-open and eyebrow-raise etc.

The 68 FAPs are categorized into 10 groups corresponding to different facial features such as eye, eyebrow, mouth, jaw and cheek etc. By referring the FAP group definition, the AU definition in FACS and observation on real facial expression image [23, 30], we proposed a set of PEP parameters defined in Table 1. The PEP aims to simulate the representative and meaningful emotional facial expression that commonly occurs in our daily life under the hypothesis that there exist fixed facial motion patterns to express specific emotional state.

Table 1. PEP definition with partial expression description

Face Region	PEP code (Left/Right)	Partial Expression Description	
		[0,-1]	[0,1]
Eye-brow	1.1(L/R)	Eyebrow lower down	Eyebrow raise up
	1.2(L/R)	Relax Eyebrow	Squeeze Eyebrow
	1.3(L/R)	In the shape of "\ /"	In the shape of "/ \"
Eye	2.1(L/R)	Close eye-lid	Open eye-lid
	2.2(L/R)	(Eyeball) look right	(Eyeball) look left
	2.3(L/R)	(Eyeball) look up	(Eyeball) look down
Mouth	3.1	Close mouth	Open mouth
	3.2	Mouth-corner bent down	Mouth-corner bent up
	3.3	Mouth sipped	Mouth protruded (pout)
	3.4	Mouth stretched in	Mouth stretched out
Jaw	4.1	Jaw move up	Jaw lower down
	4.2	Jaw move right	Jaw move left

For each PEP parameter, we defined its corresponding FAP subgroup, its direction of motion in positive value, its motion unit, and its translation mapping function to FAPs. The FAP subgroup is selected based on the FAP distribution within local regions. Most of PEP parameters are bidirectional which indicates the referred motion pattern can occur in two opposite directions to display different facial expression. For example, the mouth-corner-bent-up usually indicates happiness or joyful while mouth-corner-bent-down occurs together with sadness, anger etc. The PEP parameter, with a scalar value ranging from -1 to +1, simulates the motion intensity or degree of muscle contraction, and thus describes the continuous or subtle change of local facial expression. For better understanding, the movement of mouth-bent (PEP3.2) is illustrated in Figure 3 with its value ranging from -1 to +1. The motion unit of PEP parameter is designed based on the motion unit of FAP (FAPU). In the following section, we introduce the PEP-FAP translation template which convert PEP parameter to corresponding FAPs.

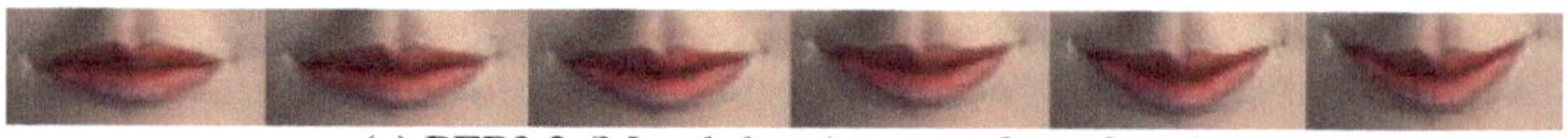

(a) PEP3.2 (Mouth-bent) ranges from 0 to 1

(b) PEP3.2 (Mouth-bent) ranges from 0 to -1

Fig. 3. Partial expression movement of mouth-bent ranging in [-1, 1

4.2 PEP-FAP Translation Template

For each PEP parameter, a subgroup of FAP is defined for animating face model to perform corresponding local facial expression described by the PEP. Previous study on FAPs correlation indicates that there exist high correlations among different FAPs, which has been successfully exploited to interpolate the unknown FAP value from a set of decided FAPs [26]. This motivates us to implement a PEP-FAP translation template to convert the PEP to its corresponding FAPs.

According to Lava's study on FAP correlation [26], three sets of FAPs are analyzed which correspond to three main facial features, namely eyebrow, eye, and mouth, all of them are the most active and expressive facial features to display facial expression. The selection of correlated FAPs is based on a priori knowledge about the anatomy of human faces. Previous research defined key FAP as the FAP that has the highest sum of correlation coefficients with respect to other FAPs, and the other FAPs are interpolated by the key FAP [26].

Table 2. Related FAP group and key FAP for each PEP parameter

Face Region	PEP Code	FAP Number	Key FAP	Related FAP Group
Eye-brow	1.1	6	F33	[F33,F31,F35,F34,F32,F36]
	1.2	2	F37	[F37,F38]
	1.3	6	F31	[F31,F35, F33, F32, F36, F34]
Eye	2.1	4	F19	[F19,F21,F20,F22]
	2.2	2	F23	[F23,F24]
	2.3	2	F25	[F25,F26]
Mouth	3.1	12	F5	[F5,F10,F11,F52,F57,F58, F4,F8,F9,F51,F55,F56]
	3.2	4	F12	[F12,F13,F59,F60]
	3.3	2	F16	[F16,F17]
	3.4	4	F6	[F6,F7,F53,F54]
Jaw	4.1	1	F3	[F3]
	4.2	1	F15	[F15]

For each PEP parameter, we define its associated FAP group and key FAP as shown in Table 2. The selection of key-FAP is in accordance with the result in [26], and the other non-key FAP are linear interpolated by key-FAP. The mathematical form of the linear PEP-FAP translation template is defined in Equation 1. For the i-th PEP parameter in region R (P_i^R), the value of key-FAP (F_k) is determined by P_i^R directly, and the value of non-key FAP (F_j) is linear interpolated by the F_k with the coefficient α_k^j The F_k^{max} is the maximum value of F_k. F_k^{max} and α_k^j are experimentally determined using a home-grown facial expression editor and also referring the result in previous study [41].

$$\begin{cases} F_k = P_i^R \bullet F_k^{\max} & (P_i^R \in [-1,+1]) \\ F_j = \alpha_k^j \bullet F_k & (\alpha_k^j \in [-1,+1],\ k \neq j) \end{cases} \tag{1}$$

5 Pseudo Facial Expression Database

In order to model the relationship between PAD emotional dimensions and PEP parameters, a pseudo facial expression database is created. Here, the "*pseudo*" means that the database is not the real human facial expression but the synthetic facial expression on 3D talking avatar as shown in Figure 4.

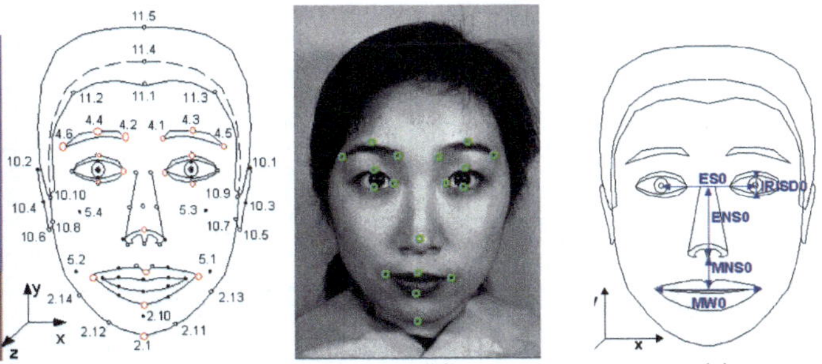

(a) (b) (c)

Fig. 4. Avatar front view **Fig. 5.** Annotated facial points (b), FDPs (a) and FAPU (c)

We choose the Japanese Female Facial Expression (JAFFE) database as reference to create our pseudo database [30]. First the PEP parameter is extracted by measuring the facial features movement in real human image, and then is used to animate talking avatar to reproduce the similar facial expression. The pseudo facial expression database can be viewed online [62]. The establishment of pseudo facial expression database validates the effectiveness of PEP on synthesizing expressive facial movement on one hand. On the other hand, the individual (such as different age, face shape, and neutral state face) and appearance (such as the skin texture) effects on facial expression perception are eliminated by reproducing facial expression on a single talking avatar. Currently in this chapter, we focus on the geometric change of facial features such as the motion of eyebrow and lip etc., while the appearance change is not considered, such as the wrinkle, furrows and laugher line etc.

5.1 JAFFE Expression Database

The JAFFE expression database contains 213 expression images with 10 Japanese females posing three or four examples for each of seven basic expressions, namely *Neutral*, *Happy*, *Sad*, *Surprise*, *Angry*, *Disgust* and *Fear*. For each image 18 facial feature points are annotated manually in accordance to MPEG-4 facial definition points (FDPs) [34] as shown in Figure 5(a) and 5(b). The PEP parameters are then extracted based on these annotations to create pseudo facial expression database.

5.2 PEP Extraction for Pseudo Facial Expression Database

Based on the above annotation, 12 PEP parameters are extracted by measuring the movement of facial feature points defined in Table 3. The other 6 PEPs are not extracted by lack of depth information (PEP 2.2 and 2.3 for eye-ball movement, PEP 3.3 for mouth movement in z-direction), or the movement is not obvious (PEP 4.2 for horizontal jaw movement) in the database. The FAPU shown in Figure 5(c) is utilized to normalize PEP to eliminate the individual difference in face shape. By applying the extracted PEP in talking avatar animation, we create the pseudo facial expression database with similar synthetic expressions as real human.

Table 3. PEP measurement by facial point movement

Face Region	PEP code	PEP Measurement	Units (FAPU)
Eye-brow	1.1(L/R)	4.3.y (L)	ENS
		4.4.y (R)	
	1.2(L/R)	4.1.x (L)	ES
		4.2.x (R)	
	1.3(L/R)	$\arctan\left(\frac{4.1.y-4.5.y}{4.1.x-4.5.x}\right)$ (L)	AU
		$\arctan\left(\frac{4.2.y-4.6.y}{4.2.x-4.6.x}\right)$ (R)	
Eye	2.1(L/R)	3.9.y – 3.13.y (L)	IRISD
		3.10.y – 3.14.y (R)	
	2.2(L/R)	not-extracted	AU
	2.3(L/R)	not-extracted	AU
Mouth	3.1	8.1.y – 8.2.y	MNS
	3.2	$\frac{8.3.y+8.4.y}{2}-\frac{8.1.y+8.2.y}{2}$	MNS
	3.3	not-extracted	MNS
	3.4	8.3.x – 8.4.x	MW
Jaw	4.1	2.1.y	MNS
	4.2	2.1.x (not-extracted)	MW

A *k-means clustering* experiment is conducted on the pseudo database to validate the reproduced facial expressions. The clustering result is presented in Table 4, which states the number of seven intended expression in each PEP clustering group. The synthetic expression of each clustering center is illustrated in Figure 6. According to the clustering result, the extracted PEP can distinguish the basic emotions to some extent in that the PEP for the same expression is clustered into the same group with the exception of *Angry* and *Disgust*. This may be reasonable because that some similar facial movements are shared by *Angry* and *Disgust* as reported in [58].

Table 4. Sample number of 7 expressions (labels in the first row) in each PEP clustering center (C_i in the first column)

	HAP	SUP	ANG	NEU	SAD	FEAR	DIS	Total
C1	**25**	1	2	0	1	0	0	29
C2	4	**20**	1	0	0	3	1	29
C3	2	0	**18**	0	9	4	***12***	45
C4	0	6	1	**30**	0	0	0	37
C5	0	0	0	0	**15**	9	2	26
C6	0	3	0	0	4	**11**	2	20
C7	0	0	8	0	2	5	**12**	27
Total	31	30	30	30	31	32	29	213

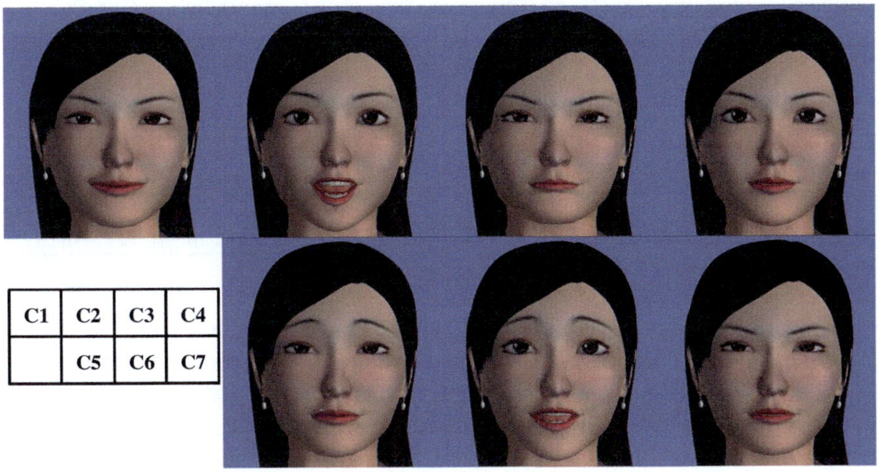

Fig. 6. Synthetic expressions for PEP value of clustering center in Table 4

6 PAD-PEP Mapping Model

The PAD emotional model proposed by Mehrabin is not only a tool to describe the human emotional state, but also enables researcher to build quantitative and clear connections between high-level human emotion perception and low-level acoustic/visual signals [4, 9, 60]. As the layered framework of PAD-driven facial expression synthesis stated in section 3, the PAD-PEP mapping model is learned from pseudo facial expression database. The database established in section 5 already contains the PEP configurations, and thus the PAD values for each synthetic expression need to be annotated for training the PAD-PEP mapping model.

6.1 PAD Annotation for Pseudo Expression Database

Based on the PAD scales proposed by Mehrabian [33], a Chinese-English bilingual version of abbreviated PAD emotion scales is provided [28], where the *P*, *A* and *D* values can be evaluated by the 12-item questionnaire shown in Table 5. The abbreviated PAD scale was proved as a versatile psychological measuring instrument that is capable of adapting to a variety of applications including emotion annotation.

Table 5. 12-item PAD questionnaire for expression annotation and evaluation

Emotion		-4	-3	-2	-1	0	1	2	3	4	Emotion	
Angry	愤怒的										Activated	有活力的
Wide-awake	清醒的										Sleepy	困倦的
Controlled	被控的										Controlling	主控的
Friendly	友好的										Scornful	轻蔑的
Calm	平静的										Excited	激动的
Dominant	支配的										Submissive	顺从的
Cruel	残忍的										Joyful	高兴的
Interested	感兴趣的										Relaxed	放松的
Guided	被引导的										Autonomou	自主的
Excited	兴奋的										Enraged	激怒的
Relaxed	放松的										Hopeful	充满希望的
Influential	有影响力										Influenced	被影响的

For each synthetic expression in the database, we annotate the *P*, *A* and *D* value using the 12-item questionnaire. During annotation, the annotator is required to describe the synthetic expression using 12 pairs of emotional words as given in Table5. For each pair of the emotional words, which are just like two ends of a scale, the annotator is asked choose the one that better describes the synthetic expression with a 9 level score varying from -4 to +4. The *P*, *A* and *D* values are then calculated from this questionnaire using the method described in [33]. Before annotating, the annotator is trained by an expert to use the PAD questionnaire.

The annotation result is summarized in Table 6, with the corresponding distribution in PAD emotional space shown in Figure 7. We can see that the PAD annotations for each basic emotion category are distributed in nearly different

Table 6. Average PAD annotation for the pseudo expression database

Intent Emotion	MEAN		
	P	A	D
ANG	-0.59	0.08	0.47
DIS	-0.59	-0.01	0.40
FEAR	-0.08	0.18	-0.39
HAP	0.63	0.40	0.29
NEU	0.03	-0.04	-0.07
SAD	-0.28	-0.12	-0.37
SUP	0.41	0.55	0.19

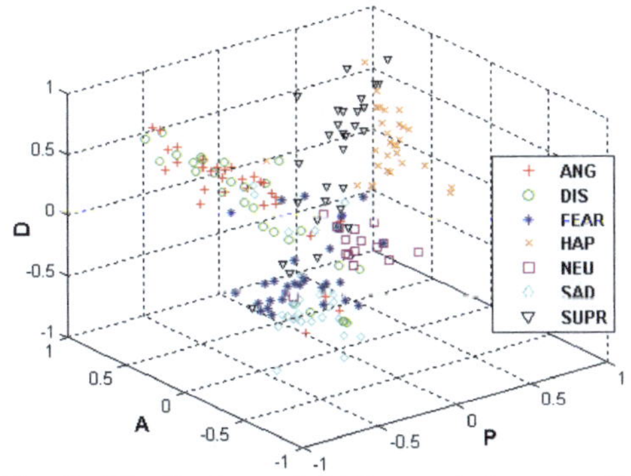

Fig. 7. Distribution of PAD annotations for the pseudo expression database

areas with the exception of *Angry* and *Disgust*. This result is consistent with the clustering result of PEP as stated in section 5.2.

6.2 PAD-PEP Mapping Model

Different from the previous work on realistic-looking facial animation, our work focuses on exploring the relationship between high-level emotion description and mid-level facial expression configuration, in other words, our goal is to implement a mapping between PAD and PEP. The pseudo expression database, which contains both the PAD annotation and PEP configuration, is used to train the PAD-PEP mapping model.

In order to find a proper emotion-expression mapping between PAD and PEP, we have tried the polynomial function with the first and second order as well as non-linear exponential function. As the experimental result reveals, it is the second order polynomial function that achieves the best fitting result, and its mathematic form is shown in Equation 2.

$$\boldsymbol{PEP} = \alpha \bullet \boldsymbol{E}^2 + \beta \bullet \boldsymbol{E} + \delta \tag{2}$$

Where ***PEP*** is the PEP vector of expression configuration, and ***E*** is the PAD vector $[P, A, D]$, $\boldsymbol{E}^2$ is a vector with each element the square of its counterpart in ***E*** respectively, i.e. $[P^2, A^2, D^2]$, $\boldsymbol{\alpha}$ and $\boldsymbol{\beta}$ are the corresponding coefficient matrix, $\boldsymbol{\delta}$ is the constant offset vector. Each dimension of ***PEP*** vector is estimated respectively with the same mathematic form as equation 3 shown. The PEP_i is the i-th dimension in ***PEP*** vector, and all the corresponding coefficients are indicated with the subscript i and P,A and D labels.

$$PEP_i = [\alpha_{Pi}, \alpha_{Ai}, \alpha_{Di}] \begin{bmatrix} P^2 \\ A^2 \\ D^2 \end{bmatrix} + [\beta_{Pi}, \beta_{Ai}, \beta_{Di}] \begin{bmatrix} P \\ A \\ D \end{bmatrix} + \delta_i \tag{3}$$

7 Integration to Emotional Text to Audio Visual Speech System

The emotional facial expression synthesis for talking avatar is a critical aspect for implementing the emotional text-to-audio-visual-speech system. Previous solutions on synthesizing emotional facial expression accompanying speech are achieved by combination of a set of basic facial expressions corresponding to different full-blown emotions [1, 5, 31, 49]. However we aims to build an emotional text-to-audio-visual-speech (ETTAVS) system based on emotion dimensions, namely the dimensions of pleasure, arousal and dominance, which has been successfully exploited in audio visual speech synthesis [9, 59, 60, 63]. This motivates us to integrate the PAD based emotional facial expression synthesis into the ETTAVS system. The overview of the system is shown in Figure 8. The input text message in first semi-automatically annotated with PAD values. The PAD value is then used to generate the emotional facial expression (i.e. PEP configuration), and to convert the synthetic speech generated by TTS engine to emotional speech [59, 60]. The emotional facial expression is then dynamically interpolated according to the speech prosody (see details in Section 8). After that, the synthetic head and facial movement are combined with the synthetic viseme generated by viseme synthesizer [54] to animate the 3D talking avatar, the audio-visual synchronization module controls the timing of audio speech play and visual speech animation [58].

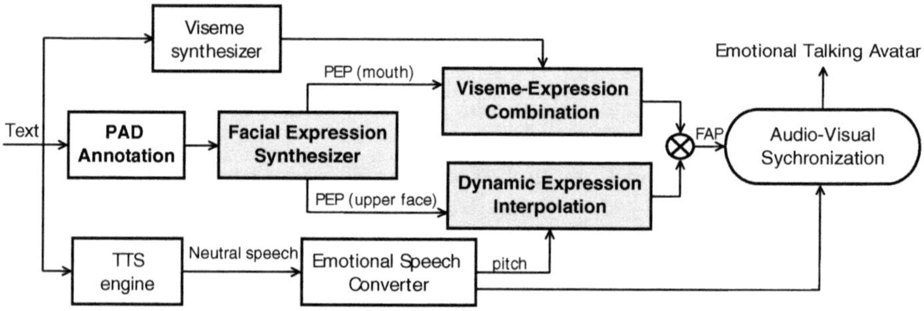

Fig. 8. Overview of the Emotional Text to Audio Visual Speech System

8 Experiments and Results

8.1 Experiments on PAD-PEP Mapping Function

The pseudo expression database consists of 213 expression samples, which are extracted from the JAFFE database consisting of 10 subjects and 7 expressions. In order to reduce the limitation caused by the small database, we adopt the K-fold cross-validation method to train the PAD-PEP mapping function. For the test set, we select 10*2=20 expression image samples by randomly selecting 2 different expressions for all 10 subjects. The training set, which consists of the rest 193 expression image samples, is then divided into 10 subsets with each subset covering all 10 subjects and each subject with 1 or 2 expression samples. By such division scheme, we are able to capture the common facial expression movement shared by different people as much as possible.

The coefficients of the PAD-PEP mapping function is estimated using the least square errors method, and each dimension of PEP (e.g. PEP_i) is estimated separately with the same function as shown in equation 3, which means that we finally have 12 sets of coefficients corresponding to 12 PEP parameters. In the K-fold cross-validation training process (k=10), there are 10 iterations corresponding to 10 validating subset. In each iteration, we calculate the correlation coefficient between real and estimate data on the validating subset. The correlation is taken as criteria to evaluate the fitting performance of the trained function. The minimum, maximum and average value of correlation coefficient is summarized in Table 7. The trained function with the average fitting performance among all the 10 iterations is chosen as the final result and used to evaluate the performance on test set.

Table 7. Correlation coefficients for evaluating PAD-PEP mapping function

PEP code	Validation set (k=10)			Test set
	Min	Max	Avg	
1.1L	0.56	0.88	0.77	0.74
1.1R	0.52	0.86	0.77	0.77
1.2L	0.13	0.82	0.43	0.51
1.2R	0.13	0.64	0.44	0.62
1.3L	0.63	0.89	0.79	0.80
1.3R	0.56	0.91	0.77	0.82
2.1L	0.53	0.84	0.75	0.60
2.1R	0.51	0.85	0.76	0.62
3.1	0.36	0.89	0.65	0.52
3.2	0.24	0.93	0.71	0.75
3.4	0.06	0.77	0.52	0.78
4.1	0.36	0.83	0.57	0.66
Average	0.38	0.84	0.66	0.69

From the result, we can see that the performance of mapping function on test set is acceptable with correlation coefficient be around 0.70. There is still some space for performance to be improved. This may be explained as that the pseudo expression database consist only 10 subjects with 7 basic expressions, which is insufficient for training the mapping function to capture common expression patterns among different people. The possible solution is either to increase the number of the subjects or to collect more expression images for a specific subject.

8.2 Synthetic Expression and Perceptual Evaluation

A series of perceptual evaluation is conducted to evaluate the performance of the layered framework for expression synthesis. The synthetic expressions for the 6 basic emotion words (*happy*, *surprise*, *sad*, *scared*, *angry*, and *disgusting*) and some other candidate emotion words with corresponding PAD value are shown in Figure 9. 14 subjects are then invited in the evaluation to finish the 12-item PAD

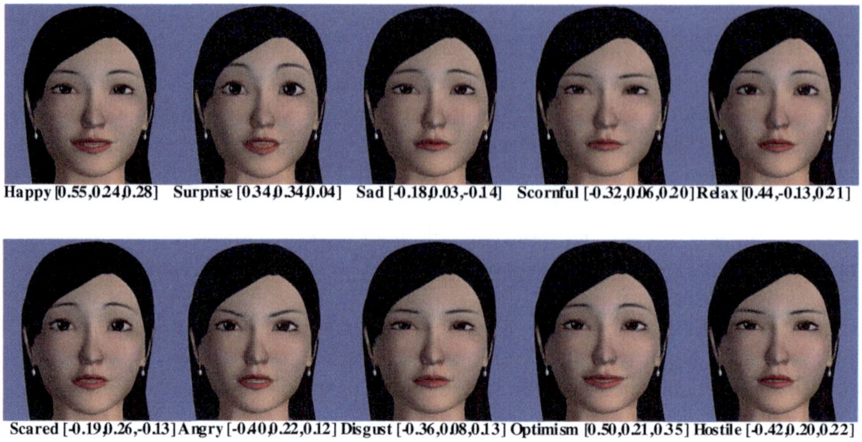

Fig. 9. PAD-driven synthetic expression for selected emotions (Emotion [P,A,D])

questionnaire as shown in Table 5 for each synthetic facial expression image. Another emotion labeling experiment is also conducted where the subjects are required to select one word that best describe the synthetic expression from 14 candidate emotion words, which are *Happy*, *Optimism*, *Relax*, *Surprise*, *Mildness*, *Dependent*, *Bored*, *Sad*, *Scared*, *Anxious*, *Scornful*, *Disgusting*, *Angry*, and *Hostile*. The PAD values for each basic emotion word and the PAD perception values for synthetic expression are summarized in Table 8.

Table 8. Result of PAD evaluation and emotion labeling. (The PAD value of *Hostile* is [-0.42, 0.20, 0.22])

PAD-driven synthetic expression	Original PAD of emotion words			Evaluation PAD of Synthetic image			Expression Label (with voting percent)
	P	A	D	P	A	D	
Happy	0.55	0.24	0.28	0.42	0.12	0.10	Happy (67%)
Surprise	0.34	0.34	0.04	0.36	0.45	-0.05	Surprise (100%)
Sad	-0.18	0.03	-0.14	-0.01	-0.26	-0.27	Sad (42%)
Scared	-0.19	0.26	-0.13	0.01	-0.04	-0.25	Sad (50%)
Angry	-0.40	0.22	0.12	-0.17	0.02	-0.08	Hostile (58%)
Disgust	-0.36	0.08	0.13	-0.56	0.15	0.44	Disgusting (50%)

A voting method is applied to determine the emotion category for each synthetic facial expression image based on the result of expression labeling. The PAD correlation coefficients between emotion words and synthetic expression are 0.89(P), 0.68(A) and 0.70(D) respectively, this result is consistent with the reliability and validity of Chinese version of abbreviated PAD emotion scales reported in [28]. Confusion between *Sad* and *Scared* as well as *Angry* and *Hostile* is found in expression labeling experiment, it is because that the input PAD value for each of the two confused words are very close, which originates from the confusion in human understanding between these emotion words and thus leads to the similarity between synthetic expressions.

The experimental results indicates that the PAD emotional parameters can be used in describing the emotion state as well as facial expression; and the proposed layered framework is effective for PAD-driven facial expression synthesis.

8.3 Dynamic Facial Expression Accompanying Speech

As stated in Section 7, the emotional facial expression is integrated into our ETTAVS system. To enhance the emotional expressivity of facial expression accompanying speech, the synthetic emotional facial expression by PAD-PEP-FAP approach is dynamically interpolated in accordance with the prosodic change in emotional speech on sentence level. Based on the observation of the video of real speaker's dynamic facial expression in a speech, we proposed the "*peak-intensify*" rule that the facial expression will be intensified when the speech prosody get aroused (i.e. the peak time). The speech pitch (F0) is utilized as important cues to locate the *peak* time, and the PEP value for facial expression at peak time is calculated as Equation (4), while the dynamics of facial expression over whole sentence is obtained by interpolating the PEP from zero to the value at peak time using the Piecewise Cubic Hermite Interpolating method [18]. Figure 10 illustrates the interpolation result of eyebrow squeeze (PEP1.2) for a sentence with disgust emotion. Figure 11 presents the synthetic facial expression series and speech signal.

$$PEP^{peak} = \alpha^{C} \bullet PEP^{main} \quad \left(\alpha = \left|F_0^{peak} / F_0^{average}\right|\right) \tag{4}$$

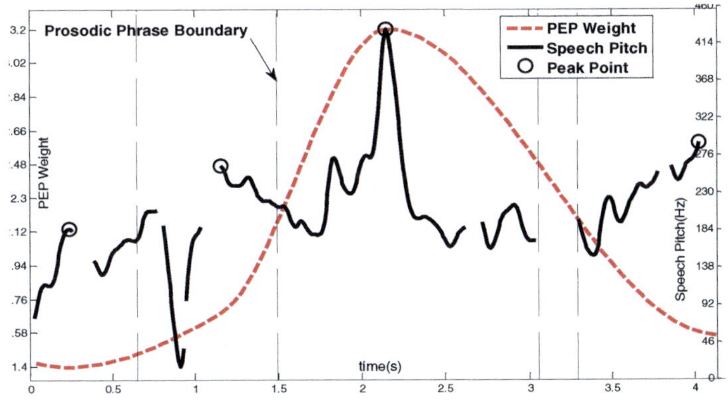

Fig. 10. PEP Interpolation based on Speech Pitch(F0) for a sentence

The *peak-intensify* rule is only applied to the upper face area, while the mouth area is not affected by this rule since the shape of mouth must be consistent with the viseme configuration. For the mouth animation in emotional speech, we take a combination of viseme and facial expression. A linear weighted function is

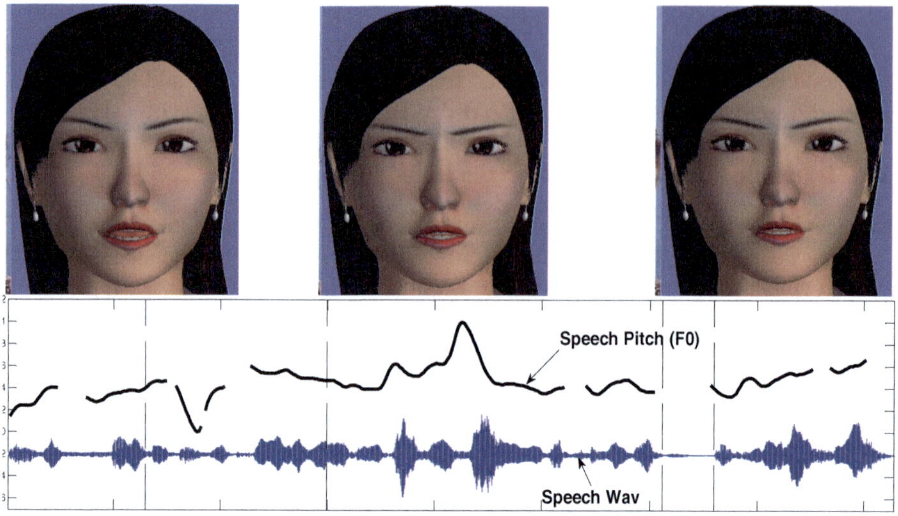

Fig. 11. Dynamics of emotional facial expression and speech prosody in a sentence

applied to merge the animation parameter (FAP) from viseme and facial expression, as shown in Equation 5. It should be noticed that, the merge function is only applied to the FAPs that control the width and height of outer lip corner, and the selection of these FAPs is experimentally manually decided to achieve best animation result. The coefficient α is manually set to determine whether the viseme dominates the lip movement or the facial expression. In this work, we take α=0.6 to obtain the final animation parameter for mouth.

$$\boldsymbol{FAP}_{\text{final}}^{\text{mouth}} = \alpha \boldsymbol{FAP}_{\text{viseme}}^{\text{mouth}} + (1-\alpha)\boldsymbol{FAP}_{\text{expr}}^{\text{mouth}} \tag{5}$$

9 Conclusion and Further work

This chapter introduces our work on synthesizing emotional facial expression on expressive talking avatar in emotional text to audio visual speech system. A layered framework of parametric facial expression synthesis based on emotion dimensions (i.e. pleasure-arousal-dominant) is proposed. The facial expression is synthesized according to the high-level PAD emotional parameters, which indicates the intended or desired emotional state of the talking avatar. The mid-level partial expression parameter (PEP) is designed to depict the facial expression movement within a specific face region (i.e. eyebrow, eye and mouth etc.). The MPEG-4 facial animation parameter (FAP) is adopted as low-level parameter for direct manipulation of facial points on the avatar. The PAD-PEP mapping model and PEP-FAP translation template are then implemented to translate the PAD parameters to PEP parameters and then to FAP parameters for facial expression animation.

The proposed approach is effective for facial expression synthesis, and enhances the emotional expressivity of talking avatar. Together with our previous work on expressive head movement synthesis [63], the emotional facial animation is integrated into the emotional text to audio visual speech system (ETTAVS). A dynamic interpolation rule and a linear combination strategy are proposed to achieve more natural facial animation accompanying speech.

Further work will focus on the semantic facial expression of talking avatar that helps to deliver the communicative message in human conversation. Our next goal is to implement a conversational engaging talking avatar for spoken dialog system.

Acknowledgments. This work is supported by National Natural Science Foundation of China (60805008,90820304) ,the National Basic Research Program of China (973 Program) (No. 2006CB303101) and the National HighTechnology Research and Development Program("863"Program) of China (No. 2007AA01Z198). This work is affiliated with the Microsoft-CUHK Joint Laboratory for Human-centric Computing and Interface Technologies. We also thank Sirui Wang from Institute of Psychology, Chinese Academy of Science for providing the 14 typical emotion words and corresponding PAD values in the perceptual evaluation.

References

[1] Albrecht, I., Haber, J., Seidel, H.P.: Automatic generation of non-verbal facial expressions from speech. In: Proc. Computer Graphics International 2002, pp. 283–293 (2002)

[2] Albrecht, I., Schröder, M., Haber, J., Seidel, H.P.: Mixed feelings: expression of non-basic emotions in a muscle-based talking head. Virtual Reality 8(4), 201–212 (2005)

[3] Busso, C., Deng, Z., Grimm, M., Neumann, U., Narayanan, S.: Rigid head motion in expressive speech animation: analysis and synthesis. IEEE Transactions on Audio, Speech, and Language Processing 15(3), 1075–1086 (2007)

[4] Cao, J., Wang, H., Hu, P., Miao, J.: PAD model based facial expression analysis. In: Bebis, G., Boyle, R., Parvin, B., Koracin, D., Remagnino, P., Porikli, F., Peters, J., Klosowski, J., Arns, L., Chun, Y.K., Rhyne, T.-M., Monroe, L. (eds.) ISVC 2008, Part II. LNCS, vol. 5359, pp. 450–459. Springer, Heidelberg (2008)

[5] Cao, Y., Tien, W.C., Faloutsos, P., Pighin, F.: Expressive speech-driven facial animation. ACM Trans. on Graph 24 (2005)

[6] Cohn, J., Zlochower, A., Lien, J., Kanade, T.: Automated face analysis by feature point tracking has high concurrent validity with manual FACS coding. Psychophysiology 36, 35–43 (1999)

[7] Cowie, R., Douglas-Cowie, E., Tsapatsoulis, N., Votsis, G., Kollias, S., Fellenz, W., Taylor, J.G.: Emotion recognition in human-computer interaction, vol. 18(1), pp. 32–80 (2001)

[8] Cowie, R., Douglas-Cowie, E., Savvidou, S., McMahon, E., Sawey, M., Schröder, M.: FEELTRACE: an instrument for recording perceived emotion in real time. In: Proceedings of the ISCA workshop on speech and emotion, Northern Ireland, pp. 19–24 (2000)

[9] Cui, D., Meng, F., Cai, L., Sun, L.: Affect related acoustic features of speech and their modification. In: Paiva, A.C.R., Prada, R., Picard, R.W. (eds.) ACII 2007. LNCS, vol. 4738, pp. 776–777. Springer, Heidelberg (2007)

[10] Darwin, C.: The expression of the emotions in man and animals. University of Chicago Press, Chicago (1965)
[11] Du, Y., Lin, X.: Emotional facial expression model building. Pattern Recognition Letters 24(16), 2923–2934 (2003)
[12] Ekman, P.: Universals and cultural differences in facial expressions of emotion. In: Cole, J. (ed.) Proc. Nebraska Symposium on Motivation, vol. 19, pp. 207–283 (1971)
[13] Ekman, P.: About brows: emotional and conversational signals. In: Human ethology: claims and limits of a new discipline: contributions to the Colloquium, pp. 169–248. Cambridge University Press, England (1979)
[14] Ekman, P., Friesen, W.: Facial action coding system: A technique for the measurement of facial movement. Tech. rep. Consulting Psychologists Press (1978)
[15] Faigin, G.: The Artist's Complete Guide to Facial Expression. Watson-Guptill (2008)
[16] Fasel, B., Luttin, J.: Recognition of asymmetric facial action unit activities and intensities. In: Proceedings of International Conference of Pattern Recognition (2000)
[17] Friesen, W., Ekman, P.: Emfacs-7: emotional facial action coding system, Unpublished manuscript, University of California at San Francisco (1983)
[18] Fritsch, F.N., Carlson, R.E.: Monotone piecewise cubic interpolation. SIAM Journal on Numerical Analysis 17, 238–246 (1980)
[19] Granstrom, B., House, D.: Audiovisual representation of prosody in expressive speech communication. Speech Communication 46(3-4), 473–484 (2005)
[20] Hong, P., Wen, Z., Huang, T.S.: Real-time speech-driven face animation with expressions using neural networks, vol. 13(4), pp. 916–927 (2002)
[21] Ibanez, J., Aylett, R., Ruiz-Rodarte, R.: Storytelling in virtual environments from a virtual guide perspective. Virtual Reality 7, 30–42 (2003)
[22] Kalra, P., Mangili, A., Magnenat-Thalmann, N., Thalmann, D.: Simulation of facial muscle actions based on rational free form deformations. In: Proc. Eurographics 1992, pp. 59–69 (1992)
[23] Kanade, T., Cohn, J.F., Tian, Y.: Comprehensive database for facial expression analysis. In: Proc. Fourth IEEE International Conference on Automatic Face and Gesture Recognition, pp. 46–53 (2000)
[24] Kshirsagar, S., Escher, M., Sannier, G., Magnenat-Thalmann, N.: Multimodal animation system based on the mpeg-4 standard. In: Proceedings Multimedia Modelling 1999, pp. 21–25 (1999)
[25] Lang, P.J.: Behavioral treatment and bio-behavioral assessment: computer applications. In: Technology in mental health care delivery systems, pp. 119–137. Ablex, Norwood (1980)
[26] Lavagetto, F., Pockaj, R.: An efficient use of mpeg-4 fap interpolation for facial animation at 70 bits/frame, vol. 11(10), pp. 1085–1097 (2001)
[27] Li, S.Z., Jain, A.K.: Handbook of Facial Recognition. Springer, New York (2005)
[28] Li, X., Zhou, H., Song, S., Ran, T., Fu, X.: The reliability and validity of the chinese version of abbreviated PAD emotion scales. In: Tao, J., Tan, T., Picard, R.W. (eds.) ACII 2005. LNCS, vol. 3784, pp. 513–518. Springer, Heidelberg (2005)
[29] Linden Research, Inc.: Second life: Online 3D virtual world, http://secondlife.com/
[30] Lyons, M., Akamatsu, S., Kamachi, M., Gyoba, J.: Coding facial expressions with gabor wavelets. In: Proc. Third IEEE International Conference on Automatic Face and Gesture Recognition, pp. 200–205 (1998)
[31] Mana, N., Pianesi, F.: Hmm-based synthesis of emotional facial expressions during speech in synthetic talking heads. In: Proceeding of 8th International Conference on Multimodal Interfaces (ICMI 2006), Banff, AB, Canada, pp. 380–387 (2006)
[32] Mehrabian, A.: Communication without words. Psychology Today 2, 53–56 (1968)

[33] Mehrabian, A.: Pleasure-arousal-dominance: A general framework for describing and measuring individual differences in temperament. Current Psychology: Developmental, Learning, Personality, Social 14(4), 261–292 (1996)
[34] Motion Pictures Expert Group: ISO/IEC 14496-2.: International standard, information technology-coding of audio-visual objects. part 2: Visual; amendment 1: Visual extensions (1999/Amd. 1: 2000(E))
[35] Mower, E., Mataric, M.J., Narayanan, S.: Human perception of audio-visual synthetic character emotion expression in the presence of ambiguous and conflicting information. IEEE Transaction on Multimedia 11(5), 843–855 (2009)
[36] Oddcast Inc.: Personalized speaking avatars service, http://www.voki.com/
[37] Osgood, C.E., Suci, G.J., Tannenbaum, P.H.: The Measurement of Meaning. University of Illinois Press (1957)
[38] Parke, F.I.: Parameterized models for facial animation, vol. 2(9), pp. 61–68 (1982)
[39] Ekman, P., Rosenberg, E.L.: What the face reveals: basic and applied studies of spontaneous expression using the facial action coding system (FACS). Oxford University Press, US (2005)
[40] Picard, R.W.: Affective Computing. MIT Press, Cambridge (1997)
[41] Raouzaiou, A., Tsapatsoulis, N., Karpouzis, K., Kollias, S.: Parameterized facial expression synthesis based on MPEG-4. EURASIP Journal on Applied Signal Processing 2002, 1021–1038 (2002)
[42] Reallusion, Inc.: Crazytalk for skype, http://www.reallusion.com/crazytalk4skype/
[43] Plutchik, R.: A general psychoevolutionary theory of emotion. In: Emotion: Theory, research, and experience. Theories of emotion, vol. 1, pp. 3–33. Academic, New York (1980)
[44] Russell, J., Mehrabian, A.: Evidence for a three-factor theory of emotions. Journal of Research in Personality 11, 273–294 (1977)
[45] Russell, J.A., Fernez-Dols, J.M. (eds.): The Psychology of Facial Expression. Cambridge University Press, Cambridge (1997)
[46] Ruttkay, Z., Noot, H., Hagen, P.: Emotion disc and emotion squares: Tools to explore the facial expression space. Computer Graphics Forum 22(1), 49–53 (2003)
[47] Schröder, M.: Dimensional emotion representation as a basis for speech synthesis with non-extreme emotions. In: André, E., Dybkjær, L., Minker, W., Heisterkamp, P. (eds.) ADS 2004. LNCS (LNAI), vol. 3068, pp. 209–220. Springer, Heidelberg (2004)
[48] Schröder, M.: Expressing degree of activation in synthetic speech. IEEE Transactions on Audio, Speech, and Language Processing 14(4), 1128–1136 (2006)
[49] Tang, H., Huang, T.S.: MPEG4 performance-driven avatar via robust facial motion tracking. In: International Conference on Image Processing, ICIP, San Diego, CA, United state, pp. 249–252 (2008)
[50] Terzopolous, D., Waters, K.: Physically-based facial modeling, analysis and animation. Journal of Visualization and Computer Animation 1, 73–80 (1990)
[51] Theune, M., Meijs, K., Heylen, D., Ordelman, R.: Generating expressive speech for storytelling applications. IEEE Transactions on Audio, Speech, and Language Processing 14(4), 1137–1144 (2006)
[52] Tian, Y.I., Kanade, T., Cohn, J.F.: Recognizing action units for facial expression analysis, vol. 23(2), pp. 97–115 (2001)
[53] Tsapatsoulis, N., Raousaiou, A., Kollias, S., Cowie, R., Douglas-Cowie, E.: Emotion recognition and synthesis based on MPEG-4 FAPs. In: MPEG-4 facial animation-the standard implementations applications, pp. 141–167. Wiley, Hillsdale (2002)
[54] Wang, Z., Cai, L., AI, H.: A dynamic viseme model for personalizing a talking head. In: Sixth International Conference on Signal Processing (ICSP 2002), pp. 26–30 (2002)

[55] Waters, K.: A muscle model of animating three dimensional facial expression. Computer Graphics 22(4), 17–24 (1987)

[56] Welbergen, H., Nijholt, A., Reidsma, D., Zwiers, J.: Presenting in virtual worlds: Towards an architecture for a 3D presenter explaining 2D-presented information. In: Maybury, M., Stock, O., Wahlster, W. (eds.) INTETAIN 2005. LNCS (LNAI), vol. 3814, pp. 203–212. Springer, Heidelberg (2005)

[57] Whissell, C.: The Dictionary of Affect in Language Emotion: Theory, Research and Experience. In: The Measurement of Emotions, vol. 4, pp. 113–131. Academic Press, London (1989)

[58] Wu, Z., Zhang, S., Cai, L., Meng, H.M.: Real-time synthesis of Chinese visual speech and facial expressions using mpeg-4 fap features in a three-dimensional avatar. In: INTERSPEECH 2006 and 9th International Conference on Spoken Language Processing, vol. 4, pp. 1802–1805 (2006)

[59] Yang, H., Meng, H.M., Cai, L.: Modeling the acoustic correlates of expressive elements in text genres for expressive text-to-speech synthesis. In: INTERSPEECH 2006 and 9th International Conference on Spoken Language Processing, vol. 4, pp. 1806–1809 (2006)

[60] Yang, H., Meng, H.M., Wu, Z., Cai, L.: Modelling the global acoustic correlates of expressivity for Chinese text-to-speech synthesis. In: Proc. IEEE Spoken Language Technology Workshop, pp. 138–141 (2006)

[61] Zeng, Z., Pantic, M., Roisman, G.I., Huang, T.S.: A survey of affect recognition methods: Audio, visual, and spontaneous expressions. IEEE Transaction on Multimedia 31(1), 39–58 (2009)

[62] Zhang, S.: Pseudo facial expression database, `http://hcsi.cs.tsinghua.edu.cn/Demo/jaffe/emot/index.php`

[63] Zhang, S., Wu, Z., Meng, H.M., Cai, L.: Head movement synthesis based on semantic and prosodic features for a Chinese expressive avatar. In: Proc. IEEE International Conference on Acoustics, Speech and Signal Processing ICASSP 2007, vol. 4, pp. 837–840 (2007)

Chapter 7

Attentive Conversational Agent with Internal State Transition for Multiple Users

Hiroki Ohashi, Hung-Hsuan Huang, and Toyoaki Nishida

Kyoto University, Yoshida-Honmachi Sakyo-ku Kyoto 606-8501, Japan

Abstract. In this chapter, we discuss how to add emotion to virtual agents and how this influences human-agent communication. Firstly, we conducted a preliminary experiment to investigate which kind of emotions we human beings represent during human to human communication. How do we represent them, and how they change. As a consequence of this analysis, we constructed a state transition model, where the agent has emotions as internal states. We change them according to the particular situation. To realize this model, we thought that the method of machine learning and especially the method of Support Vector Machine(SVM), would be useful. We implemented an agent using that model, and conducted an experiment. We did questionnaire survey and a GNAT test to evaluate the performance of the implemented agent. After analyzing these results combined together, we concluded that our model worked well when representing the natural nonverbal behaviors, which should convey agent's present emotion. We couldn't get a satisfactory result in the aspect of verbal behavior. Further research is necessary especially on the verbal aspect. In addition to this, we also plan to consider a model of users' emotion during the conversation and then integrate it into the model.

1 Introduction

Embodied Conversational Agents (ECAs) are life-like virtual characters that are able to engage in face-to-face conversations with humans. Since ECAs have the ability to display verbal and nonverbal communication like human beings, they can be expected to play an important role as alternatives to humans. This ability can provide valuable support to humans. They can provide an interface between humans and computers. One of the most important things in such situations is their ability to simulate life-likeness. Improving those attribute can provide the users with a sense of immersion. This makes the whole system more attractive.

The present ECAs don't have a sufficient degree of naturalness and life-likeness because there are many difficulties to achieve this. They are very ambiguous and it is difficult to construct a model which simulates these qualities well. Since Humans must sense the meanings of information during conversations and use them for determining their own behaviors, ECAs can't avoid the difficulties of sensing such information correctly. Furthermore, even

T. Nishida et al. (Eds.): Modeling Machine Emotions for Realizing Intelligence, SIST 1, pp. 133–155.
springerlink.com

if the very good model and very good sensors were given, the problem of how to represent their behaviors would be still left.

In this chapter, we propose one possible way to improve naturalness of ECAs. We pay special attention to the "emotion" of the agents. We believe this helps to improve their naturalness.

We decided to develop a Quiz Agent, which subjects the users to a quiz game. We chose the quiz agent as a target application for the following reasons. There may be many conversations among the users during quiz games. The behavior of the users can often change during quiz games. The agent can have many chances to contribute by speaking to the users or by changing its behavior. We think that these situations help us to find what is important in human-agent communication.

In the following sections, we first look at related works and especially those concerned with the agent's emotion in Sect. 2. In Sect. 3, we introduce the preliminary experiment. This serves to investigate which type of emotion we human beings represent in human to human communication. We investigate how we represent them, and how we change them. In Sect. 4, we describe how our model is obtained from the analysis of the experiment in detail. Section 5 explains how this was implemented. Sect. 6 describes details of our experiment. This was conducted to evaluate the implemented agent and the results obtained. We also consider the findings of the experiment and discuss the following: Does our model work well? What influence did our model have on the users? What is the effect on human-agent communication when the ECAs have an emotional factor? Finally, we describe our future works and propose some ideas to extend the model in Sect. 7.

2 Related Works

ECAs have been studied as one field in the research on Artificial Intelligence. They are based on many technologies in many fields. These include Speech Recognition, Speech Synthesis, Natural Language Processing and Computer Graphics [1][2][3]. At the time when the research started, researchers tried to develop ECAs with the aim of enabling general users to use computers without special knowledge. This was done by using their daily communication skill. The demands in this field have increased greatly recently. Additional qualities such as "Kindness", "Life-likeness", and "Believability" of the agents are becoming necessary. The challenge to provide Emotion to agents has begun.

There are two famous and often used theories to describe emotion. The model proposed by Ortony, Clore and Collins[4](the OCC Model) and Mehrabian's PAD(Pleasure, Arousal, and Dominance) Model[5]. The OCC model is a Cognitive Emotion Model. In the OCC model, emotion is regarded as Valenced Reactions to Events, Agents, or Objects. Each different emotion is determined as a result of evaluation of these elements. The PAD model tries to explain various kinds of emotions using a three dimensional space: PAD space. The axis P means Pleasure-Displeasure. This is defined as Positive versus Negative Affective States. The axis A denotes Arousal-Nonarousal. This is defined in terms of the level of Mental Alertness and Physical Activity. The axis D means Dominance-Submissiveness. This is defined as a feeling of Control and Influence over versus Controlled or Influenced by one's surroundings and ther people. Each emotion is described by a combination of three values of those axises. A great deal of work has been done by using these models. For example, the Virtual Human Max[6], ALMA(A Layered Model of Affect)[7] employ both of these emotional models.

Gratch and Marsella[8] discussed how psychological theory can be related to the interaction between emotion and cognition. They introduced a computational framework for modeling emotion which treats appraisal and coping as a central principle. Rapport agent[9] by Gratch et al. is also regarded as an example which considers emotion in human-agent communication. They reported that when listening to a user's speech, a responsive agent was just as effective as human lister in creating rapport. This was even although it responds without understanding the content of the speaker's speech. In [10], Ball et al. tried to express emotion using a two dimensional model; One axis is Valence and the other axis is Arousal. Based on the emotion, they succeeded in producing an intuitively plausible sequence of background behavior for an embodied agent. In [11], Wilson presented an overview of the Artificial Emotion Engine and review of the neurophysiological model used by the engine. He confirmed the validity of the model.

There have been many works on emotion and its artifacts. What is important for our particular case? To be more specific, which kind of emotion should our agent have? How should emotions be represented? How should the emotions change? We conducted a preliminary experiment in order to find answers to these questions. The next section describes the experiment and its findings.

3 A Preliminary Experiment

In this section, we describe the detail of our preliminary experiment and its findings. We then discuss which aspects of emotion play an important role especially in human to agent communication. We try to determine how we can give emotion to the various agents.

We now explain the configuration of our first experiment.

Goal. We investigate the following:

- Which kind of emotions should our agent have when giving a quiz game?
- Which kind of information is important when trying to express the emotion?
- How should the emotions be best represented?
- How should the emotions change?

Participants. Three male students at our university. Not laboratory members. These participants are friends.

Setting. This preliminary experiment was a kind of Wizard of Oz(WOZ) experiment. A WOZ experiment is a way in which participants interact with a computer system which is operated by a human. Here one of us introduced a quiz game. The presenter reported which kind of emotion he had when he was giving a quiz game, and which kind of information he paid attention to during the quiz session. In addition we also did a video analysis after the experiment.

Quizzes. We did a series of quizzes of the following kinds. Some of them are about knowledge on various things which included trivia. Others demanded inspiration or logical thinking.

Environment. We created an environment which simulates the situation that an agent uses for a quiz game(Fig.1). We prepared two rooms. Room A, where users gave answers to the quizzes; Room B, where the presenter introduces the quiz. These rooms were connected using Skype. This is a software application which can be used for video conferencing. Both the users and the presenter were able to see and listen to each other via the Skype program. This setting allowed the users see the screen which displayed the presenter and quizzes(Fig.2).

Findings

We now summarize the findings from the interview to the presenter and the video analysis.

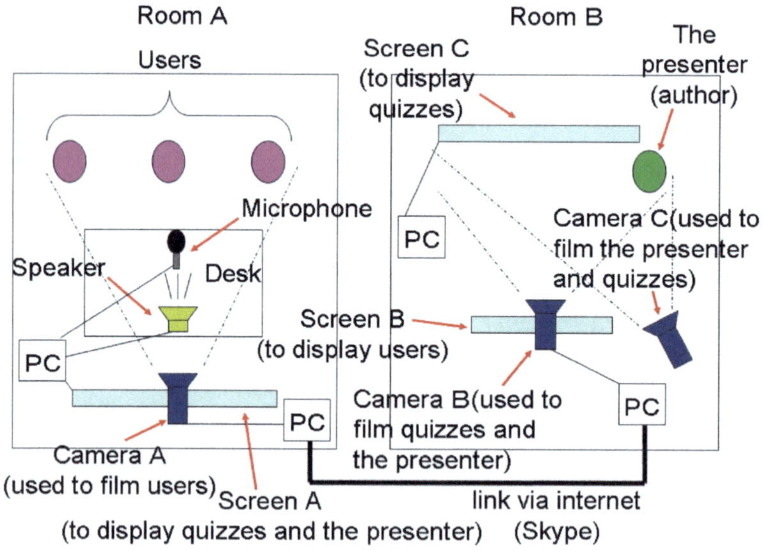

Fig. 1. The configuration for the preliminary experiment

(a) users (Camera A, Screen B) (b) Screen which users see (Camera B, Screen A) (c) the presenter (Camera C)

Fig. 2. Screen shots from each camera

1. The presenter's emotions and the situation were classified into following 3 categories:
   (a) **Calm:** Users are enjoying the quiz game and show considerable interest. They are also actively discussing the quiz. The presenter does not think it necessary to intervene. He remains quiet.
   (b) **Anxious:** Users are earnestly working on the quiz game, but they don't discuss the task with each other very much. The presenter wishes to provoke a discussion by introducing a hint.
   (c) **Impatient:** Users are not facing the screen but to one and other. They seem to be having fun by themselves. As time passes, the presenter becomes motivated and urges them to answer and continue to the next quiz.
2. The information the presenter paid attention to was as to whether there was a discussion among users, and whether the users are facing towards the screen.
3. The presenter's emotions changed according to the information on users. This information is what we mentioned above (in the item 2).
4. The emotions also changed depending on the time which had passed since he'd given the quiz and the contents of the quiz.
5. The presenter's way of representing behaviors, not only the active ones but also the passive ones, was based on his emotion at the moment.
6. The presenter spoke to the users several times. He proposed to give a hint or urging them to answer. He didn't speak to them immediately he wished, but waited for a suitable time, when the users became calmer.

Based on those findings, which model is appropriate to show the agent which exhibits emotion? The process of constructing the model is given in the next section.

4 Constructing the Model Which Is Able to Simulate the Emotion - State Transition Model

We describe how we constructed the model based on the findings in the preliminary experiment.

4.1 Requirements and System Model

Based on the findings of the preliminary experiment, the factors the agent should meet are summarized as follows:

1. The agent should see if users are discussing the quiz or not and if they are facing towards the screen or not.
2. The agent should represent the behaviors based on its emotions.
3. The agent should alter its emotions based on the information on users, the elapsed time, and the type of the quizzes.
4. The agent should mainly use three emotions such as those described in the first finding found in Sect. 3.
5. The agent should speak to users on occasions. When it tries to speaks, it should not start until the users become calm.

In response to the above requirements the following system model was adopted.

System Model
We define the internal states in the agent as representative of its emotions. The agent's behavior is based on the internal states which are transmitted according to the users' nonverbal behavior. Other information concerned with quiz game is also considered. The internal states are divided into 3 main categories. The agent occasionally speaks to the users.

4.2 The Internal State in Quiz Game Presenters

We name three states mentioned as the forth requirement in Sect. 4.1. These are *Calm, Anxious, and Impatient.* We establish two levels, Weak and Strong, in the Anxious and Impatient sections in order to meet the 5th requirement the agent must speak to the users at an appropriate time. The two levels are defined as follows:

Weak : The agent represent its internal states passively
Strong : The agent aggressively tries to choose good timing to speak to the users, when representing its internal states.

Consequently, we defined five internal states in the agent: *1. Calm, 2. Anxious weak, 3. Anxious strong, 4. Impatient weak, and 5. Impatient strong.* The five states are shown in Fig. 3. We now explain each of the states respectively and how the agent behaves in the state.

1. Calm
This state means that the quiz game is proceeding well and there is no need to try to intervene actively. The agent should also adopt this state for a certain time after introducing the quiz (Fig. 4).

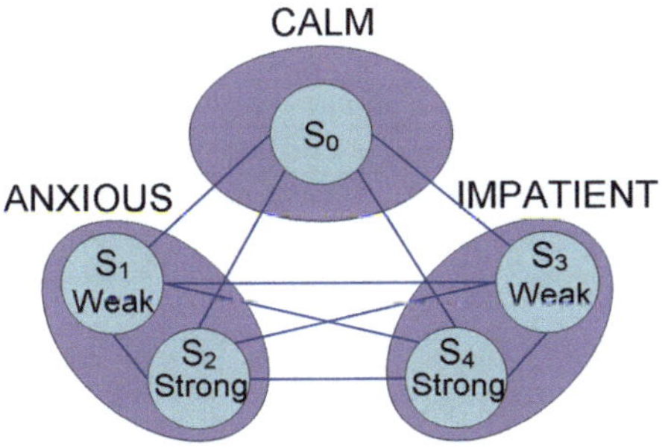

Fig. 3. State diagram

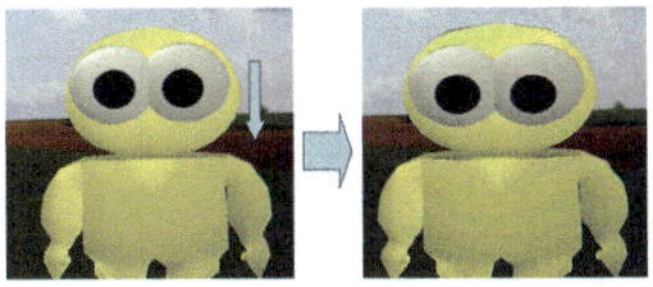

The agent nods on occasion.

It doesn't move its head, hands, and feet vigorously.

Fig. 4. Calm

2.Anxious-weak

This state means that the agent is becoming anxious about the users (Fig.5). The agent looks at the users one by one because it is worrying about them. The typical situation is that users appear to have an interest in the quiz, but don't discuss it actively and consequently are not excited.

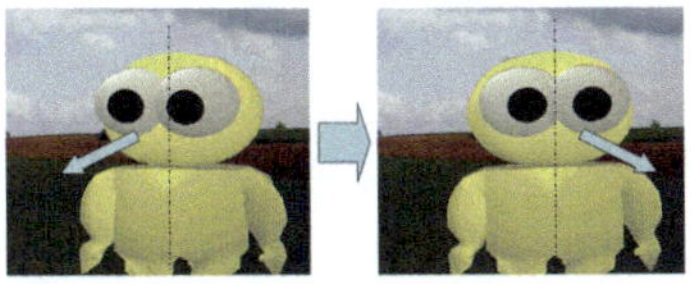

The agent is looking at the users.

It doesn't move its hands and feet significantly.

Fig. 5. Anxious-weak

3.Anxious-strong

This state means that the agent is very concerned about the users (Fig.6). Perhaps this is because they don't discuss the quiz. It seems that the users lost an interest to the quiz. The agent tries to intervene in the users so that the users talk more. A possible strategy is that it tries to provoke discussion by proposing to give a hint.

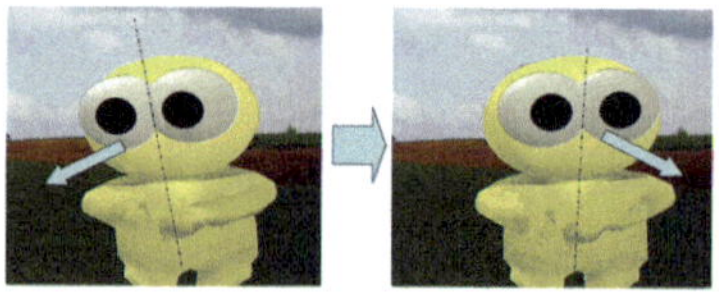

The agent is paying attention to the users.
This is shown by bending its body forward.
It's rubbing its hands together.
It's very anxious about the users.

Fig. 6. Anxious-strong

4.Impatient-weak
This state means that the agent is a little irritated with users(Fig.7). A typical situation is that users keep chatting and pay no attention to the agent or the quiz despite the elapsing of time.

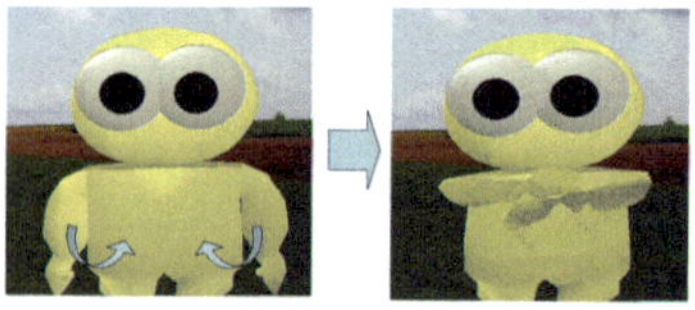

The agent is folding its arms.
The feet don't move.

Fig. 7. Impatient-weak

5.Impatient-strong
This state means that the agent is very impatient with users (Fig.8). The agent aggressively tries either to encourage the users to find a conclusion or to make a decision.

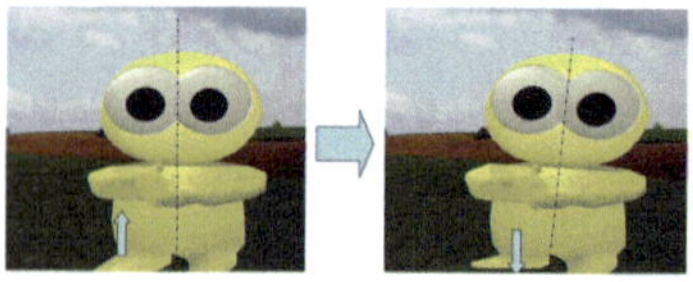

The agent is folding its arms.
It moves its feet up and down restlessly
and standing as a slant.
It's very impatient with the users.

Fig. 8. Impatient-strong

4.3 The Rule of Transition

When we try to use nonverbal information to calculate the agents' state, the number of the input combinations is numerous. It is difficult to detect the rule of state transition by using statistical analysis. This is a lengthy process. A method of machine learning was adopted. We chose the method of Support Vector Machine. This is widely studied and reported to be stable to use.

The Design of SVM
We now describe how to design SVM and examine the input, output, and the kernel in this subsection.

Input
We chose the following inputs according to the requirements mentioned in Sect. 4.1. These include;

1. The voice volume
2. The direction of users' face
3. Elapsed time since the quiz was given
4. Kind of the quiz:
   - Knowledge: requires the knowledge of various things such as history, geography, or science
   - Logic: requires an ability to think logically. This includes deduction for example
   - Wit: requires intelligence

Output
The output of the SVM is the current internal state of the agent. To be specific, which of those defined in Sect. 4.2. It means that the output is within *Calm, Anxious-weak, Anxious-strong, Impatient-weak, or Impatient-strong.*

Kernel
We used the Radial Basis Function(RBF). The RBF kernel maps data into a higher dimension nonlinearly. It was expected to be suitable in this instance, as it seems to be impossible to separate the data linearly. It is superior to other kernels such as the polynomial kernel or the sigmoid kernel due to the complexity.

How to Get the Input
We now explain how to obtain the input to the SVM.

1. The voice volume: We gather this by a microphone fixed to the table in front of the users. The collected voice is quantized using 8 bits. The sampling rate is 22.4 kHz. We apply Fast Fourier Transform(FFT) to each 1024 data. The noise is removed. The logarithm of the spectrum is summed up. We use this as the volume of the users' voice.
2. The direction of users' face: We use OKAO Vision[12], a commercial software produced by OMRON Corporation for determining the direction of the users' face. Okao means a face in Japanese. Okao Vision is software which contains technologies in these areas: Face Detection, Facial Feature Extraction, Face Recognition, Facial Attribute Estimation, Automatic Optimum Facial Picture Adjustment. We decided to use only the Face Direction Detection(Fig.9). We can get stable and reliable data in the real world environment. The score of users' face direction is calculated using the information on face direction detection. It is obtained in the form of angle of pan, tilt, and rotation. These are represented in degrees. The reliability of the result is an integer between 0 and 1000.

Fig. 9. Detection of the face direction using OKAO Vision

3. Elapsed time: Every time a quiz was given, the counter is reset to be zero, and the elapsed time from then can be calculated. It is represented in an integer.
4. Kind of the quiz: The kind of information was knowledge, logic, or wit. The pending quiz is categorized and represented in the form 0/1. If the kind of the quiz is "logic", then the values are {0, 1, 0}.

How to Obtain the Training Data

We explain how we created the training data for the SVM. We used iCorpus Studio[13] which was developed in our laboratory to analyze and create the data. iCorpusStudio is software which enables us to easily analyze various data simultaneously. It can synchronize various sets of data such as video, sound, motion capturing, and physiological indices. It also enables us to label this data. We can do some operations on those labels. We watched the synchronized videos and sounds of the users and the person making the presentation by using iCorpusStudio. We then added information on the presenter's state. We analyzed only the time between the presenter gave the quiz and when the users answered. We calculated the volume of the voice, the face direction, the elapsed time, and details of the quiz, by using the algorithm explained above. The training data was obtained.

The total time of the quiz session was 11 minutes and 17 seconds, and 8 minutes and 16 seconds was annotated. In this time the training data was created every 33.32msec. The number sets of training data was 14890.

In the following subsection, we summarize the constructed classifier.

4.4 The Constructed Classifier

When we implemented the SVM, we used LIBSVM[14]. This is a library for SVM created and made available to the public by Chih-Chung Chang, and Chih-Jen Lin, Taiwan university.

We applied a two class classification to all the pairs out of the five classes. It means that we built $10({}_5C_2)$ sets of one-versus-one classifiers. After applying all of these classifications, we chose the class that was selected by the most classifiers. The values of the threshold θ of each 2 classes are shown in Table 1. The numbers of the support vectors(SV) are shown in Table 2.

We used the RBF kernel as we have stated above. The value of γ in the RBF kernel was set at 0.5.

Table 1. The threshold of each 2 classes

Class	1 - 2	1 - 3	1 - 4	1 - 5	2 - 3
Threshold	-2.75925	2.07495	4.01747	-1.04463	-10.2751
Class	2 - 4	2 - 5	3 - 4	3 - 5	4 - 5
Threshold	-1.9772	-2.47827	11.1825	8.44725	-3.84704

(1:Calm,2:Anxious-weak,3:Anxious-strong,4:Impatient-weak,5:Impatient-strong)

Table 2. The number of support vectors

Class	1	2	3	4	5	Total
Number of SV	2062	2246	1190	1757	883	8138

We conducted a 10-fold cross validation and achieved about 73.2% in classification rate.

5 Implementation

In this section, we introduce how we implemented our agent. Firstly, we give the overview of the architecture in Sect. 5.1. We briefly then describe each of the components in the following subsections.

5.1 System Architecture

Our system is based on GECA(Generic Embodied Conversational Agent) Framework[15], which was developed in our laboratory. It is a platform for developing the ECA system and it can support various components necessary for developing ECAs. By using it, we can develop and revise each component without concern for the other components. This is irrespetive as to where they are located or whether they were modified or not.

We present the system architecture of our implementation shown in Fig.10. The raw data are obtained through the use of camera, microphone, and touch panel, and these are processed at the appropriate component. Then, the processed data is gathered in the SVM component, where the current internal state of the agent is calculated. The SVM component sends the current state to the scenario component. The scenario component manages the information involved during the process of the quiz game, It sends the information on

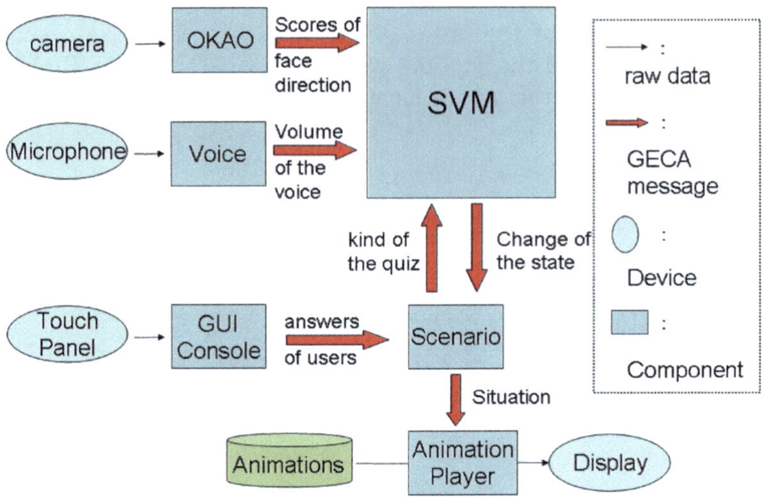

Fig. 10. System architecture

the state of the quiz game to the animation player, where the animation will be played in accordance to the output of the SVM component. The player component interprets the information from the scenario component, and it plays an appropriate animation.

5.2 GUI Console Component

This component is used to receive the answer to the quiz obtained from the users. It senses the users' touch on the touch panel, and judges the choice of the users. It then sends the results to the scenario component.

5.3 Voice Component

This component is used to calculate the volume of the voice. It gathers the audio information of the users, and processes it as we mentioned in Sect. 4.3. Then, it sends the calculated volume to the SVM component about 8 times per second.

5.4 OKAO Component

This component is used to calculate the scores of the face direction of the users. It gathers the visual information of the users, and processes it as we mentioned in Sect. 4.3. Then, it sends the calculated scores to the SVM component about 4 times per second.

5.5 SVM Component

This component is used to classify the agent's internal state using the SVM classifier mentioned in Sect. 4.3. It calculates the internal state of the agent about 3 times per second.

5.6 Scenario Component

This component is used to judge which animation is to be played and which utterance is to be given. This refers to the process of the quiz game which is managed by this component. The users' answers are obtained via the GUI console component. The agent's internal states are obtained via the SVM component. The animations decided and the utterances are sent to the animation player component. If the information from GUI console component and the information from the SVM component collide, this component decides which should have priority.

5.7 Animation Player Component

This component is used to play animations which are determined in the scenario component. We use Visage|SDK^{TM}[16]. This is a commercial product of Visage Technologies AB. It is a C++ Software Development Kit for virtual character animation. It provides extensive support for both interactive and off-line animation of multiple characters with fully integrated real time lip-sync, speech synthesis, facial features detection and tracking (facial motion capture), body animation and much more. It is based on the MPEG-4 Face and Body Animation (MPEG-4 FBA). We developed an animation player using this SDK. This component accepts XML messages and plays animations.

6 Experiment

We conducted a user experiment in order to evaluate our system. We asked participants to engage in the two quiz games. One is hosted by an agent which has a model which we constructed. This is hereinafter called the *attentive agent*. The other is hosted by an agent which randomly changes its internal states. This is called the *random-state agent*. In this experiment, we used two method for evaluation. One is Normal Questionnaire Survey, and the other is the GNAT test[17].

6.1 Settings

Participants
The profiles of the participants were as follows.

Occupation Students of Kyoto university

Condition of attendance Groups of 3 people
Number of participants 24. A total of 8 groups
Gender Male:18, Female:6
Average age of the participants 22,5 years

Environment
We carried out this experiment in two rooms(Fig.11): one for the quiz game (Fig.12(a)), and one for the GNAT test(Fig.12(b)). This is to prevent disturbance between the groups.

Time. approximately 1.5 hours.
Rewards. Payment for participating the experiment was a fixed amount. Some additional rewards such as stationery, USB memory, etc. were also privided. This was related to the quiz game score.
Hardware. There were 10 PCs, 4 for the quiz game, 3 for the GNAT test, 3 for recording videos, screen, touch panel, speakers, microphone, Webcam for OKAO Vision, 3 video cameras for recording the agent, participants, and for the participants at another angle.
Agents. Two agents which are distinguished from each other by the colors used. We named them as *GECA-maru*(Fig.13(a)), and *VISA-suke* (Fig.13(b)). *"Maru"* and *"suke"* is the word which sometimes appears in the name of characters in Japan. The pairs (color and name) are fixed.

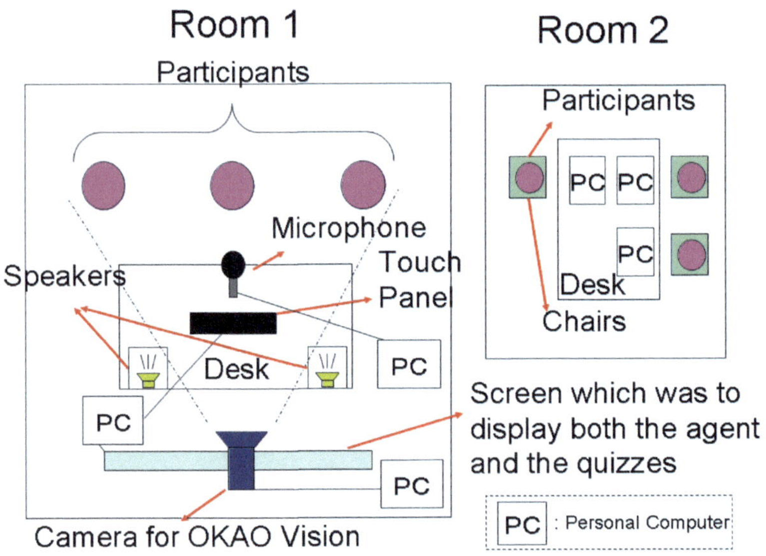

Fig. 11. The configuration of the experiment

(a) Room 1 (b) Room 2

Fig. 12. The settings in the rooms

(a) GECA-maru (b) VISA-suke

Fig. 13. They can be distinguished by their colors. GECA-maru is yellow, and VISA-suke is pink.

Quizzes. We asked the participants to engage in 2 sets of quiz games. Where one of the sets was composed of 10 quizzes.

Combination of algorithms, agents(colors), and quiz sets. 8 groups of participants engaged in one combination out of $8(= 2^3)$ possible combinations respectively. Which one to work on first in each choice :{attentive agent or random-state agent}, {GECA-maru or VISA-suke}, {Quiz set A or Quiz set B}.

Process of the Experiment

We will now show the process of the experiment.

1. Explanation of the task to the participants. The process was explained and the participants were informed of the requirements. They were also informed about the reward system. The higher the score in the quiz game the bigger the reward.
2. The operation of the quiz game (1st set).
3. The questionnaire (about the 1st set of the quiz game).
4. The operation of the quiz game (2nd set).
5. The questionnaire (about 2nd set of the quiz game).

6. The explanation of the GNAT test and how to do it.
7. The operation of the GNAT test. Details are found in Sect. 6.2.
8. Office procedure. Participants sign the agreement and filled in the form in order to obtain their payment.
9. Awarding of prizes. We awarded some prizes in accordance with the scores obtained.

6.2 The GNAT Test

What is GNAT?
The GNAT test is a method of indexing an implicit attitude or belief by assessing the strength of the association between a Target Category and two poles of an Attribute dimension. It is based on Signal Detection Theory (SDT). The strength of association is assessed by the degree to which items belonging to the Target Category and Attribute can be discriminated from Distractor items that do not belong to those concepts. For example, let the Target Category be fruit and two poles of an Attribute dimension be good and bad. Fruit is generally likely to be regarded positively. In this case, the hypothesis is that accuracy in discriminating fruit and good items from Distractors ought to be higher than accuracy in discriminating fruit and bad items from Distractors. The difference in accuracy (or sensitivity in SDT's terminology) between these conditions is taken as a measure of automatic attitude.

How to execute a GNAT test?
The test procedure is as the followings. The Test Category term (e.g. fruit) and the Test Attribute (e.g. good) is shown at the left-upper and the right-upper corners of the program window(Fig.14). The subject has to categorize a word or a picture (stimuli) shown at the center coincides with either the Category or the Attribute. The stimulus may be Targets (Signal, e.g. orange, delicious) or Distractors (Noise, e.g. carrot, rotten) and are shown on the screen for a very short interval. The response deadline is usually from 500 to 1,000 ms. If the subject's judgment is positive, then he or she has to press the space key within the response deadline(Go), or do nothing otherwise(No-go). If the judgment is correct, a green "o" will be shown at the bottom of the screen and a red "x" will be shown otherwise. This is during the time between two trials. This is known as the Inter Stimulus Interval(ISI).

Practice trials are done so that the subjects can learn the correct categorizations. A correct "Go" is called a "Hit" and an incorrect "Go" is called a "False Alarm". The sensitivity d' of a subject to the Test Category and the Test Attribute is defined as $d' = Z(h) - Z(f)$. Where h is the ratio of the number of Hits over all Signal trials and f is the ratio of the number of False Alarms over all Noise trials. From the definition of d', we know it is a measure of distance (unit: standard deviation) of the probabilistic distribution. This is for the user to have positive reaction to a Signal stimuli from the normalized noise distribution. Where the value is 0, the subject is assumed to have

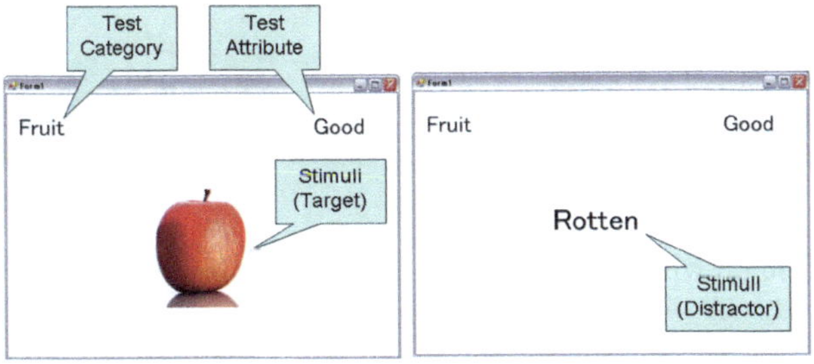

Fig. 14. An example of a GNAT test

pressed the key randomly. If the value is negative, the subject is assumed to have misunderstood the GNAT procedure. Only positive values should be considered when examining the GNAT results.

The Setting for Our Experiment

In our experiment, we have tried to investigate which of the agents, the Attentive Agent or the Random-State Agent, is more strongly associated with the concept "Natural". The stimuli used are shown in Fig.15, Fig.16, and Table 3. Fig.17 is the screen captures for our GNAT test.

(a) (b) (c) (d) (e)

Fig. 15. The stimuli which are categorized as GECA-maru. They are also used as the Distractors of VISA-suke. His body is yellow.

The Result of GNAT Test and Discussion

The result is shown in Table 4. Table 5 shows the mean value and the variance of d'.

Table 6 gives the result of the t-test. Here a null hypothesis indicates that there is no difference between the mean values of d' of each agent. The invalid samples, which have either a 0 or negative value of d' are removed. This table represents that the null hypothesis is rejected at a 1% significance level. We can statistically insist that our attentive agent has higher association with the concept "Natural" than a random-state agent. As a result, it can be safely said that our agent has succeeded in evoking feeling of "Natural".

Fig. 16. The stimuli which are categorized as VISA-suke. They are also used as the Distractors of GECA-maru. His body is pink.

Table 3. Targets and Distractors of "Natural"

Target	Human-like	Attentive	Thoughtful	Reasonable
Distractor	Awkward	Jerky	Artificial	Mechanical
Target	Consistent	Expedient	Adequate	—
Distractor	Fake	Strange	Weird	—

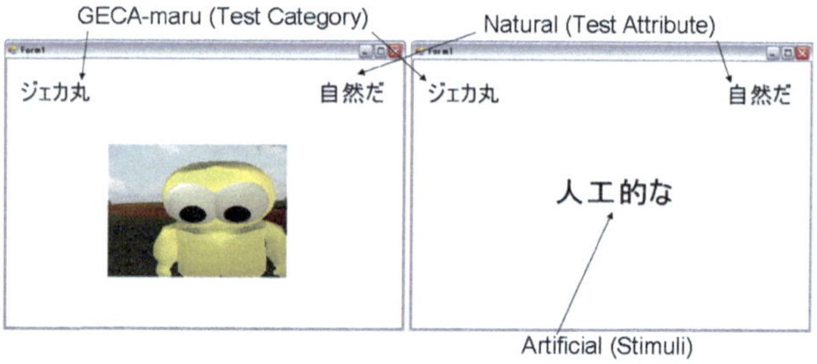

Fig. 17. Screen captures for our GNAT test. The test was carried out in Japanese.

6.3 The Questionnaire Survey

We asked each participant to answer the questionnaire immediately after they had finished one quiz session. Table 7 shows the results. The type of the questionnaire was 7 point Likert scale, where 1 means "strongly disagree" and 7 means "strongly agree". Since the results represent ordinal scale level, we conducted the Wilcoxon signed-rank test, which is a non-parametric test, for each question.

We now consider the table in detail.

Firstly, there is only one thing which can be said statistically. That is according to number 10, the game progress managed by our agent was not smooth. The question is what caused this result. In order to investigate this result, we will consider some other questions. From numbers 3 and 4, we find the participants felt that the attentive agent did not speak very aggressively. It can be interpreted to have both a good and a bad meaning. If interpreted in a good meaning, the attentive agent was not so annoying to the participants

Table 4. The result of GNAT test. The numbers in the table are the value of d' in GNAT test. '—' means Not a Number. The results are grouped according to the participants' group.

Number	Attentive	Random-state	Number	Attentive	Random-state
1	1.199	0.128	13	2.030	1.290
2	0.253	0	14	0.651	0.757
3	1.049	1.060	15	0.928	0.379
4	1.028	1.407	16	1.806	—
5	0.928	1.060	17	—	1.120
6	1.227	1.561	18	1.199	1.036
7	1.645	0.456	19	—	-0.842
8	1.561	0.967	20	0	0.800
9	2.073	0.524	21	-0.399	0.716
10	1.227	0.511	22	1.519	0.911
11	0.911	0.910	23	0.911	0.896
12	1.227	0.800	24	1.036	0.639

Table 5. Mean value and Variance of d'

	Attentive	Random-state
Mean value	1.242	0.850
Variance	0.150	0.137

Table 6. Values in t-test

Pearson Correlation	t-statistic	t Critical two-tail	P(T≤t) two-tail
$-1.002\ 10^{-2}$	3.089	2.898	<.01

as the random-state agent was. However, it also implies that the attentive agent should have spoken to the participants more actively. We think this might have affected the result of number 10. Our agent's progress was felt to be less smooth. This is possibly because our agent had less utterance. We believe the amount of the utterance should influence the progress of the quiz game.

We see number 5, 6, and 11 next. They don't have very strong meaning statistically, but still we can find out some meanings from them. If we pay attention to the numbers of "+" and "-" of these questions, we will easily notice that the participants probably felt that the attentive agent was worse than the random-state agent in these items. The common thing among these three question is that all of them are involved with utterance. This implies that utterance strategy of the attentive agent was not appropriate.

Finally, we consider the rest. Numbers 1, 2, 7, 8 and 9 show that there was no significant difference between the attentive agent and the random-state agent in those questions. What is common among those questions is that they

Table 7. The result of the questionnaire. The "+" column means the number of the participants who rated the attentive quiz agent higher than the random-state quiz agent. The columns "-", "0" mean the number of the participants who rated the attentive agent lower than, the same as the random-state agent respectively. The "p" column shows the two-tailed probabilities of Wilcoxon signed-rank test.

No.	Question	+	-	0	p
1	The character was friendly	8	8	8	0.49
2	The character was human-like	8	7	9	0.75
3	The character spoke, when I didn't want it to speak	8	13	3	0.48
4	The character was too passive	10	6	8	0.21
5	The character's utterances were appropriate	5	9	10	0.61
6	The timings of the character's utterances were appropriate	5	13	6	0.09
7	The character behaved in responding to our status	8	10	6	0.64
8	The character's behaviors were natural	9	9	6	0.4
9	The character's behaviors were comprehensive	6	5	13	0.5
10	The game progress was smooth	2	12	10	$<.01$
11	The character's urges made me feel like respond to it	4	10	10	0.24
12	The discussion was active	4	5	15	0.42

are all involved with its way of behavior, or nonverbal aspect of the agent. This means that there was no significant difference between the behaviors of the two agent. Number 12 indicates that the attentive agent couldn't succeed in evoking discussions among participants, but didn't failed terribly.

6.4 Discussion

In this section, we consider the result of the GNAT test and questionnaire survey combined. In the GNAT test, we obtained a result that our agent is considered to be natural. In the questionnaire, we found that we must reconsider the utterance strategy. However, our model didn't fail badly in the area of nonverbal behavior. Considering these results, we can find a tendency that our agent may have obtained a good result in nonverbal behavior. In order to investigate the reason as to why there is a difference between the two methods, we examined the log of informal interviews. We did this just after the experiment. According to the interviews, some participants reported that they had better impression on the random-state agent after the quiz game. Then after all processes of the experiment had been completed, it changed and they began to feel that our attentive agent was better than the random-state agent. We have explained in Sect.6.1 that we asked participants to answer to the questionnaire on their impression of the quiz set just after they had completed the test. This may have influenced the impression of the GNAT test and the questionnaire. Why could not our agent have a good impression of the quiz game? It may be because our agent didn't engage users actively as we considered in Sect.6.3. We suppose that direct involvements with the participants may have had explicit influence on the impression. Evaluation by the GNAT test had less influence in stead, the impression through the whole processes might have affected better.

Based on the two method of evaluation, we conclude that our model helped to some extent to realize natural nonverbal behaviors in ECA. In our model, the agent's behaviors are based on the emotion of *Calm, Anxious, and Impatient*. Considering the two fact that the behaviors of the attentive agent was felt to be natural and the agent's behaviors reflect its emotion, we may succeed in conveying the emotion of the agent to the participants to some extent. We found some scenes in video logs the participants said "wow, he is irritating", or "how offensive he is!" This indicates that some of the emotional feelings of the agent is successfully communicated to the participants of those scenes.

Then, which influence did those agent's action have on the participants or on the whole situation? In the example "wow, he is irritating", the agent said "have you decided on your answer?" He exhibited an impatient behavior. The participants noticed it. It seemed they felt a necessity to hurry up evoked by the agent's verbal and nonverbal behaviors. We think this may indicate that emotion played some roll in making the communication. We suppose that as a result of giving emotion to the agent, the agent was able to express its attitude more easily. Then participants could notice its intension intuitively. As a result, the communication between the agent and the participants became more close. This should be an important role of emotion in human-computer communication.

In the latter case of examples from video logs such as "how offensive he is!", we can also find a significant role for emotion. In this case, the participants are irritated by the agent's behavior. Does this necessarily mean unacceptable behavior? We believe not. Our interpretation is that this implies participants are involved in a quiz game and deeply committed to it. Although the agent never understands what the participants spoke, we often found participants replying to the agent's utterance. They occasionally called the agent "you" as if it were a human being making some comments as to the agent's behavior, and imitating the agent's actions. These participants' reactions to the agent's behaviors are based on its emotion. They are likely to imply that the participants are both enjoying and tempted by the quiz game. Here we can confirm another important role of emotion: Emotion can help in attracting users to the application.

7 Conclusion

In this chapter, we introduced our task which was to give emotion to ECAs. We have discussed the role of emotion in human to agent communication by real examples from our agent. Firstly, we conducted a preliminary experiment to investigate which model would be most appropriate to our purpose. As a consequence of this analysis, we constructed a state transition model, in which the agent has its emotions as internal states and can change them according to the users' situation. We employed the method of SVM in this model to determine the transition rule. We then implemented an agent with that model, and conducted a user evaluation experiment. In the experiment, we applied two methods of evaluation: GNAT test and Questionnaire Survey. After analyzing these combined results, we concluded that our model worked well when representing the natural nonverbal behaviors, which could convey the agent's present emotion. We couldn't get a satisfying result where speech were involved. In the experiment, we observed some interesting reactions of the participants to the agent's emotional behaviors. We then checked the video logs several times, and reached a conclusion based on the findings. Emotion can play a significant role in making human-agent communications smooth. Emotion can help in attracting users to the application.

For future works, we need further research especially based on verbal aspects. We now judge whether the agent could speak to the participants or not by using the threshold of the voice volume. We can reconsider the threshold, or study other elements to drive the utterances. It is a possibility to consider the model again and change the rule to provoke speech. We are also planning to consider the model of the users' emotion during the conversation and integrate it into the agent's model. We have worked on a prototype of the agent which detects the users, and have finished implementing an initial prototype. This can detect the users' feelings as to whether they are enjoying themselves or not. It also can estimate who is the leader in the user group to enrich its utterance strategy. We are interested in developing this prototype more, and want to enable it to detect more emotional states of the users. We think of a model which can emulate richer human-agent communication. It refers to both the agent's emotional state and the users' emotional state. We expect that the integrated agent will provide us with information to consider emotion in human-agent communication. We would like to continue research in this area. We also hope we will have a future opportunity to introduce you other interesting findings.

References

1. Cassell, J., Churchill, E., Prevost, S., Sullivan, J., Prevost, S., Churchill, E. (eds.): Embodied conversational agents. MIT Press, Cambridge (2000)
2. Prendinger, H., Ishizuka, M. (eds.): Life-Like Characters: Tools, Affective Functions, and Applications (Cognitive Technologies). Springer, Heidelberg (2004)
3. Nishida, T. (ed.): Conversational Informatics: An Engineering Approach. John Wiley & Sons, Chichester (2007)
4. Ortony, A., Clore, G.L., Collins, A.: The cognitive structure of emotions. Cambridge university press, Cambridge (1988)
5. Mehrabian, A.: Pleasure-arousal-dominance: A general framework for describing and measuring individual differences in temperament. Current Psychology 14(4), 261–292 (1996)

6. Becker, C., Kopp, S., Pfeiffer-LeBmann, N., Wachsmuth, I.: Virtual humans growing up: From primary toward secondary emotions. KI - Kuenstliche Intelligenz 1, 23–27 (2008)
7. Gebhard, P., Schroder, M., Charfuelan, M., Endres, C., Kipp, M., Pammi, S., Rumpler, M., Turk, O.: IDEAS4 Games: Building Expressive Virtual Characters for Computer Games. In: Proceedings of the 8th international conference on Intelligent Virtual Agents, pp. 426–440. Springer, Heidelberg (2008)
8. Gratch, J., Marsella, S.: A domain-independent framework for modeling emotion. Cognitive Systems Research 5(4), 269–306 (2004)
9. Gratch, J., Wang, N., Gerten, J., Fast, E., Duffy, R.: Creating rapport with virtual agents. In: Pelachaud, C., Martin, J.-C., André, E., Chollet, G., Karpouzis, K., Pelé, D. (eds.) IVA 2007. LNCS (LNAI), vol. 4722, p. 125. Springer, Heidelberg (2007)
10. Ball, G., Breese, J.: Emotion and personality in a conversational agent. In: Embodied conversational agents, pp. 189–219 (2000)
11. Wilson, I.: The Artificial Emotion Engine, Driving Emotional Behavior. In: AAAI Spring Symposium on Artificial Intelligence and Interactive Entertainment, pp. 20–22 (2000)
12. OmronCorporation.: OKAO Vision, Website (2008), `http://www.omron.com/r_d/coretech/vision/okao.html`
13. Kijima, H., Bono, M., Sumi, Y., Nishida, T.: Development of the environment for multimodal interaction analysis [in japanese]. Information Processing Society of Japan (IPSJ) SIG technical reports, 63–70 (2007)
14. Chang, C.C., Lin, C.J.: LIBSVM: a library for support vector machines (2001), Software, available at `http://www.csie.ntu.edu.tw/~cjlin/libsvm`
15. Huang, H., Nishida, T., Cerekovic, A., Pandzic, I., Nakano, Y.: The design of a generic framework for integrating ECA components. In: Proceedings of the 7th international joint conference on Autonomous agents and multiagent systems. International Foundation for Autonomous Agents and Multiagent Systems, Richland, SC, vol. 1, pp. 128–135 (2008)
16. VisageTechnologiesAB.: visage|SDK^{TM}, Website (2009), `http://www.visagetechnologies.com/products_sdk.html`
17. Nosek, B.A., Banaji, M.R.: The go/no-go association task. In: Social Cognition, vol. 19, pp. 625–664 (2001)

Chapter 8
Integrating the Emotional Intelligence into the Virtual Technical Support Engineer

Andrey Kiselev[1], Benjamin Alexander Hacker[2], Thomas Wankerl[2], Yashimasa Ohmoto[1], Niyaz Abdikeev[3], and Toyoaki Nishida[1]

[1] Graduate School of Informatics, Kyoto University, Japan
[2] Department of Informatics, Munich University of Technology, Germany
[3] Faculty of Informatics, Plekhanov Russian Academy of Economics, Russia

Abstract. This chapter addresses the problem of modeling of emotions which is important for integrating emotional intelligence into virtual agents. In this chapter we present a virtual agent which plays a role of a technical support engineer and interacts with human user to answer questions about a complex device. We show some theoretical background and basic models of formalizing emotions along with comments about their usability in our work. We present a multi-layered system of processing of emotional information which is developed according to the modern theories and allows us to realize concurrent and mutual causal processing of different types of emotional information. We also show an evaluation methodology which we use.

Keywords: Embodied Conversational Agent, Emotional Intelligence.

1 Introduction

When dealing with any complicated device, the user is confronted with issues related to the use and maintenance of the device. These issues, for instance, can be related to such aspects as a getting started, using, maintaining and recycling the device. The service meant for providing users with all necessary information is usually called technical support.

Basically, we can define technical support as a suite of services intended for answering all the questions related to the target product. There are several possibilities to offer a technical support to users ranging from various manuals and user guides those are cheap, but not very useful, up to hot lines and call centers those are efficient, but expensive, including various interactive FAQs and HowTo guides. All of these methods can be combined in a some kind of structure for offering the most reasonable level of support depending on a product and a problem. Anyway, the essence is the same in all cases: first the problem should be figured out, then relevant information should be delivered to the user and finally feedback about usability of information and respectively about the quality of technical support should be recognized and delivered back to the system. The difference between all of this methods is the media which can range from a paper up to video conferencing. The same information delivered by different media has different

T. Nishida et al. (Eds.): Modeling Machine Emotions for Realizing Intelligence, SIST 1, pp. 157–178.
springerlink.com

effectiveness. In other words, effectiveness of information depends on media. And from the other point of view, effectiveness of media depends on the type of information. For instance, when using printed manuals users have to figure out the problem and find relevant information by their own which sometimes seems to be impossible. In that case there is also no feedback from users to company which offers a tech support. On the other hand, technical support engineer can help the user to figure out the problem, find solution and get a feedback, but this method is suitable only for non consumer products due to the high costs.

In our work we pay attention to the creation of Embodied Conversation Agents (ECAs) (Cassell, 2000), which serve as a some kind of media. Within our scenario an agent plays a role of the Virtual Technical Support Engineer communicating with human users. This allows us to get all of the advantages from each of previously mentioned kinds of media. Having abilities for storing a lot of information about device usage and problem solving virtual agent offers natural face-to-face communication along with rich abilities of information representation within a virtual environment.

But first, let us have a look at key features, abilities and skills which we expect to see in this kind of media. First, we are expecting the natural face-to-face communications. But what are natural face-to-face communications?

When communicating with other people we somehow share information between each other. But which kinds of information expressing methods do we use? And on the other hand, which abilities do we use to understand our interlocutors?

Undoubtedly, the most of the time we speak. We use knowledge in language to express our thoughts clearly. We "send" a text information to our interlocutor. This information supposed to be recognized and probably understood. But what is the role of the text information in our everyday communications? It is considered that text information is only a tenth of overall flow of information which we send to and receive from out interlocutors. Of course, speaking skills and abilities to formulate clearly our thoughts can change this index, but usually other methods of information mediating are much more efficient.

When pronouncing words we add intonations, voice pitch. Thus a simple text becomes a speech with a some kind of stylistic coloring of words. We can easily convert predicate into a question just by modifying intonation. We even can use an intonation as a main information medium. For instance, the same word in the same context pronounced with different intonation can have completely different meaning.

In addition, we use a body language, which includes facial expressions, gestures, poses. We can actively use a body language to express our thoughts as well as sometimes we can not control our body language completely and it can say much more than we want to pronounce.

Abilities and skills of using all of these kinds of communications are the cornerstone in mutual understanding between humans.

Turning back to the key features of the virtual technical support engineer, we can say that it also should be able to recognize a problem or a question, to search for relevant information, which should be stored or accessible, to get a feedback from the user. For instance, we can try to imagine situation when a printer does not work and user calls to a tech support engineer. First, engineer will ask some questions about device's LED blinking or beeps whose can happen (if device is not observable by an engineer), trying to understand the problem. This can be an initial stage of the problem solving. After that engineer can suggest to plug printer to the wall or to put a paper into a paper tray, which is a process of finding and delivering relevant information to the user. Finally, engineer can get explicit or tacit reaction form a user. And all of these steps should be done using the natural communications with human user. Simplifying the aforesaid, the agent should be quite intelligent to play a role of the technical support engineer. But what is intelligence? What does it mean to be intelligent? For instance, our machines are quite intelligent if we are talking about storing and looking for information. But obviously it is not enough to be a tech support engineer. So, how many different intelligences can we mark out? Which role plays each of many intelligences in conversations or which of all kinds of intelligences are the most important in conversations? There are many works which attempts to answer these questions and we will not refer to all of these works. In the next paragraph we would like to give a brief description of intelligence and it's components.

2 An Intelligent Agent

First of all, Intelligence is an umbrella term covering the wide field of human's abilities and skills. As in cases with some other "wide terms", there is no all over accepted definition of intelligence. But there are at least two major definitions of intelligence though which both are influenced by the Herrnstein and Murray's book "The Bell Curve" (Herrnstein & Murray, 1994) kicking off a ball in a new round of debates about the nature and measuring of intelligence.

First, came from the "Mainstream Science on Intelligence" written by psychology professor Linda Gottfredson and signed by 51 other professors and published as an opinion piece in 1994 in the Wall Street Journal, says that "Intelligence is a very general mental capability that, among other things, involves the ability to reason, plan, solve problems, think abstractly, comprehend complex ideas, learn quickly and learn from experience. It is not merely book learning, a narrow academic skill, or test-taking smarts. Rather, it reflects a broader and deeper capability for comprehending our surroundings – "catching on", "making sense" of things, or "figuring out" what to do." (Gottfredson, 1994)

The second definition was proposed by the American Psychological Association in 1995. It says that "Individuals differ from one another in their ability to understand complex ideas, to adapt effectively to the environment, to learn from experience, to engage in various forms of reasoning, to overcome obstacles by taking thought. Although these individual differences can be substantial, they are never entirely consistent: a given person's intellectual performance will vary on different occasions, in different domains, as judged by different criteria. Concepts of "intelligence" are attempts to clarify and organize this complex set of phenomena". (Neisser et al., 1998)

These two definitions give us an understanding that intelligence is the wide concept which comprehends different human's abilities, each of those contributes to overall performance.

Turning back to the application domain we can say that an artificial agent which plays a role of a virtual technical support engineer has to be enough intelligent to help users with solving problem and has to be enough intelligent to communicate with users. Here we have got at least two different types of intelligence: intelligence as information and methods of searching of relevant information for solving problems and intelligence for understanding the problem and delivering relevant information to a user. During last several decades we have already made a progress in integrating the first type of intelligence into artificial agents. We also have made a progress in the second part of work by introducing a natural way of communication between humans and agents using agent's voice and body as a media. And this means that now we are giving our artificial agent's with abilities to interact with us in the most natural (for us) way. We do not need to write some special data for an agent during conversation anymore – they can listen; we do not need to explain them the features of the object we are talking about – they can see. They can (or at least they will be able soon) interact equally with us growing up to become a part of society by gaining their social intelligence. Here we have come to the term of social intelligence.

The term of social intelligence was initially proposed by Thorndike in 1920. According to the initial definition the Social Intelligence is "the ability to understand and manage men and women, boys and girls, to act wisely in human relations" (Thorndike, 1920). In the Gardner's model of multiple intelligences (Gardner, 1983) it is equivalent to interpersonal intelligence. At the same time, turning back to the topic of this chapter, according to Salovey and Mayer who coined the term of emotional intelligence, it can be treated as "a form of social intelligence that involves the ability to monitor own and others feelings and emotions, to discriminate among them, and to use this information to guide thinking and action" (Salovey & Mayer, 1990).

Obviously, emotional intelligence being a part of interpersonal intelligence on one hand plays an essential role in generating user's behavior and on the other hand helps to understand meaning of behavior of other people. Lets take a look into the nature of the term of Emotional Intelligence.

Originally the term of emotional intelligence was mentioned by Leuner in 1966, (Leuner, 1966) however it is usually attributed to Payne's "A study of Emotions: Developing Emotional Intelligence" (Payne, 1986). Few years later Salovey and Mayer attempted to answer why some individuals are better at recognizing emotions than others where they first used the term "emotional intelligence" (Mayer, DiPaolo & Salovey, 1990). Then in 1995 Goleman published the "Emotional Intelligence: Why It Can Matter More Than IQ" where he showed weakness of traditional IQ measurement and offered the "proof" that emotional intelligence is the strongest indicator of human success. Later Bradberry and Greaves in their "Emotional Intelligence Quick Book" (Bradberry, 2005) brought the term of emotional intelligence into a business world giving a framework for usage of emotional intelligence.

There are two major approaches for modeling emotional intelligence. First one, the ability based model proposed by Salovey and Mayer, is based on the idea that emotional intelligence can be defined as the ability to perceive emotion, integrate emotion to facilitate thought, understand emotions and to regulate emotions to promote personal growth. This model employs emotions as useful sources of information that help one to make sense of and navigate the social environment. The ability based model distinguishes between four types of abilities, which are:

– perceiving emotions – the one's ability to detect and recognize others emotions and ability to detect and recognize one's own emotions;
– using emotions – the ability to employ emotions to facilitate cognitive processes;
– understanding emotions – the ability to understand relationships among different emotions and comprehend the language of emotions;
– managing emotions – the ability to influence own and others emotions.

Second model proposed by Goleman after reviewing work made by Salovey and Mayer, defines a range of skills and competencies. His model distinguishes among four domains which have nineteen categories of skills and competencies. These domains are:

– self-awareness;
– self-management;
– social awareness;
– relationship management.

According to Goleman, an important thing in this model is that EI competencies are not innate talents, but learned capabilities that must be worked on and developed to achieve outstanding performance. Goleman posits that individuals are born with a general emotional intelligence that determines their potential for learning emotional competencies. (Boyatzis, Goleman & Rhee, 2000).

Each of these two models has a measuring methods. The current measure of Mayer and Salovey's model of EI, the Mayer-Salovey-Caruso Emotional Intelligence Test (MSCEIT) is based on a series of emotion-based problem-solving items.(Salovey & Grewal, 2005) Consistent with the model's claim of EI as a type of intelligence, the test is modeled on ability-based IQ tests. By testing a person's abilities on each of the four branches of emotional intelligence, it generates scores for each of the branches as well as a total score.

The Emotional Intelligence Appraisal, which was created in 2001 and which can be taken as a self-report or 360-degree assessment (Bradberry & Greaves, 2005) is currently used as a measure based on Goleman model.

3 Emotions, Reasoning and Behavior

When talking about integrating emotional behavior into the virtual agent we should ask does emotional intelligence really play a role in communications as a kind of media or in cognitive processes as a kind of facilitator? Let us start from the second part of this question.

And the answer is... Yes. Many researches in the field of emotional intelligence attempted to proof this hypothesis and one of the strongest evidences has come from a study by Cavallo and Brienza in 2002 (Cavallo & Brienza, 2002). In this study they assessed the leadership behavior of 358 leaders at Johnson and Johnson Corporation and found positive correlation between leaders' performance and score in emotional intelligence rated by their colleagues in the Emotional Competency Inventory (ECI), a 360-degree feedback instrument based on Goleman's model. There are also some other studies which argue to this hypothesis.

Let us return to the first part of question. As we mentioned before, when communicating with others we use speech and body to express thoughts. And not only speech and language of a body are a media in this case. Our actions (or inactions) also help to represent information. In other words, when communicating with others we exhibit behavior and look at others behavior. At the same time there are studies which attempts to prove the existence of causality between emotions and behavior. There are at least two different opinions about this causality one of which says that emotions can directly cause behaviors and another says that emotions are feedbacks of behaviors, but they can cause learning processes causing indirectly further behaviors.

In this chapter we would like to refrain from the opening a dispute about these theories. Instead we would like to use a pure outcome of these theories which is important for us: since intuitively there is a some kind of causality between emotions and behaviors then emotions obviously can be exhibited by using behaviors and emotions can be observed by observing behavior. This is just a theoretical background of processes we can see in our every-day live when communicating with others.

4 Brief Description of the Theoretical Background

As we mentioned before, we would like to use an idea that emotions are parameters of behavior refraining from the disputes about the psychological definition of emotions. In this paragraph we would like to show two theories which defines features of emotions ans are relevant to our work.

Appraisal theory of emotions indicates that emotions are extracted from evaluations (appraisals) of events. Thus specific reactions for the same event under the same conditions can be generated in different people. (Scherer et al., 2001)

Theory about distinguishing between two major categories of emotions has been proposed by Antonio Damasio (Damasio, 1994). The main idea concerning this theory is that all our emotions can be divided into Primary and Secondary categories.

The highest-level neural mediator of primary emotion is the limbic system, and in particular the amygdala and cingulate gyrus. Emotional responses are usually initiated in these structures in response to percepts, and move from there to the rest of the body, evoking responses in the autonomic nervous system and the production of hormones and neurotransmitters. The evolutionary advantage of this system of emotional coordination is that recognition of a threat or opportunity in the brain results in appropriate preparatory actions in the rest of the body. The release of neurotransmitters also affects brain functions, limiting or facilitating particular types of cognition as the situation demands.

Secondary emotions are controlled primarily by the frontal cortices, which invoke emotional responses in response to thoughts rather than percepts. Instead of regulating its own independent emotional production centers, the frontal cortices send signals to the limbic system to generate an emotional response. These signals may be full-fledged emotions, or may only produce part of the response spectrum, such as neurotransmitter release.

Having a physical emotion in response to a secondary emotional stimuli does have a purely survival value, as when a threat is anticipated before it can be perceived. The impairment of secondary emotional responses can also have a dramatic effect on more abstract adaptations such as social fitness. Damasio explains this with his "somatic marker hypothesis," which posits that when a thought generates a frontal cortex emotional response, the thought is "marked" by the physical emotion. Likewise, the cognitive processing of the thought is affected by the neurotransmitters evoked by the emotion. Further processing of the thought will be inhibited by negative emotions, and facilitated by positive emotions.

This very important theory give us a possibility to distinguish between fast primary emotions and secondary emotions, allowing us to implement a difference processing of these different types of information, decreasing a lag in agent's fast primary emotional reactions.

5 Approaches to Building a Computational Model of Emotions

When integrating an emotional behavior in the architecture of a virtual agent, the one of the main questions is how exactly to describe emotional states. There were a lot of attempts to solve the problem of describing emotions and several model were proposed. In this paragraph we would like to have a look at 3 different well-known and widely used models, which are important and relevant to our work – the Ortony-Clore-Collins model, Plutchik's model and PAD emotional scales.

In 1988 Ortony, Clore and Collins proposed a model of cognitive appraisal of emotions, widely known now as an OCC model (Ortony, Clore, Collins, 1988). The foundation of this model is the idea that emotions are "valenced reactions to events, agents or objects, with their particular nature being determined by the way in which the eliciting situation is construed."

This model describes 11 pairs of opposite concepts considered to be emotions. These are: gratification – remorse, gratitude – anger, pride – shame, admiration – reproach, joy – distress, happy-for – resentment, gloating – pity, hope – fear, satisfaction – fears-confirmed, relief – disappointment, love – hate.

This model gives us the set of 22 concepts supposed to be emotions. OCC is one of the most widely used models today. There are several models which are based on the OCC model.

In 1958 Plutchik (Plutchik, Hope, 1997) proposed a cone-shaped model to describe emotions with their relationships. In the model Plutchik connected emotions to colors, so that intensity of an emotion can be synthesized by synthesizing color.

The model is based on the set of 4 pairs of primary opposite emotions: joy – sadness, anger – fear, trust – disgust and anticipation – surprise. Plutchik suggested that these emotions are biological primitives. Plutchik's circumplex model is a hybrid of both basic-complex categories and dimensional theories. It arranges emotions in concentric circles where inner circles are more basic and outer circles more complex. Notably outer circles are also formed by blending the inner circle emotions.

Pursuing the parallels among the semantic differential factors and emotions, Mehrabian and Russell (1995) proposed that pleasure-displeasure (i.e., positive versus negative affective state) was the emotional counterpart of high-low evaluation of stimuli. They suggested arousal-nonarousal (i.e., level of physical activity and/or mental alertness) as a positive correlate of stimulus activity. Finally, Mehrabian and Russell proposed dominance-submissiveness (i.e., feelings of control and influence over others and situations versus feeling controlled and influenced by external circumstances) as the negative correlate of stimulus potency.

Three preliminary measures of state pleasure-displeasure, state arousal-nonarousal, and state dominance-submissiveness were developed. Each of the measures consisted of six adjective pairs, with the adjectives in each pair separated by nine spaces. For each adjective pair, participants placed a check in one of the nine spaces to describe their feelings. Sample adjective pairs were as follows: happy-unhappy, pleased-annoyed, and contented-melancholic for state pleasure; stimulated-relaxed, jittery-dull, and excited-calm for state arousal; and controlling-controlled, influential-influenced, and important-awed for state dominance. Factor analysis of all 18 items in the three measures showed three nearly independent factors: There was a -.07 correlation between Factor 1 (Pleasure) and Factor 2 (Arousal) and a .03 correlation between Pleasure and Factor 3 (Dominance); a. 18 correlation was found between Factor 2 (Arousal) and Factor 3 (Dominance). The Pleasure, Arousal, and Dominance factors accounted for 27%, 23%, and 14% of total variance, respectively.

These three models shows different attempts to solve the problem of formalizing emotions.

6 Integrating the Processing of Emotions into the Existing Architecture

6.1 Architecture

Initially, we developed the agent which was entirely devoted for pure reasoning. The architecture of this agent is based on ideas and solutions which laid in the two different frameworks – GECA (Huang et al., 2008) and SAIBA (mindmakers.org, online) – and uses the pure BDI (Bratman, 1987) reasoning core.

A framework which has been developed to ease the inclusion of these requirements is the Generic Embodied Conversational Agent (GECA) Framework. It provides some advantages compared to previously developed architectures. Components developed with different programming languages and running on different operating systems can be easily integrated into the framework. Since the components can be distributed over a TCP/IP network those which require heavy computation can be placed on different computers and thus improve the overall system performance. The single-layer component hierarchy shortens the paths of decision making and eases the support of an agent's reactive behaviors. Synchronization of the different components is achieved by providing explicit temporal information and synchronization specifiers. Loose coupling of components allows for on-line switching and upgrading of components easily.

The SAIBA (Situation, Agent, Intention, Behavior and Animation) research group's Multimodal Behavior Generation Framework is an attempt to unify the generation of natural multimodal output for embodied conversational agents. The first step in the process of developing this framework has been to lay down the general planning stages and knowledge structures that are involved in the creation of multimodal communicative behavior. The goal is the definition of representation languages that build the interface between the processing stages on three different levels: Intent Planning, Behavior Planning and Behavior Realization.

The architecture of the agent is shown on fig. 1.

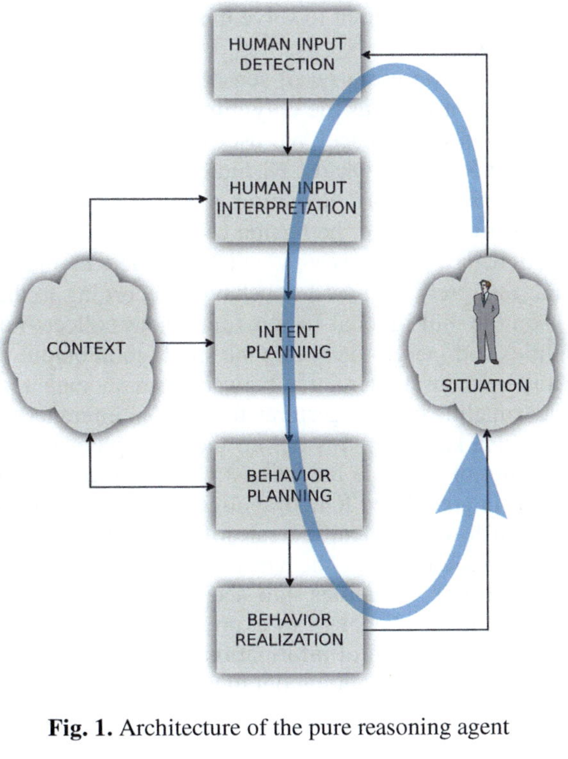

Fig. 1. Architecture of the pure reasoning agent

In the next few paragraph we would like to give description of each component of this architecture.

6.1.1 Human Behavior Detection (HBD Component)

We use a component that analyzes the behavior of user with a set of sensors and merge these data into a single output for further processing for the realization of natural-like conversation between an engineer and a virtual user.

Our system uses four types of information from the user: the movement of head and hand gestures and speech. For speech recognition we use the Automatic Speech Recognition (ASR) system developed by Loquendo Vocal Technologies (Loquendo, online) and Services. Motion data is acquired by inertial sensors developed by Ascension Technology Corporation (Ascention, online) which are attached to the body. In this setup head and single arm motion is detected by these sensors. To gain hand gesture data we use a data glove which can detect the flexed angles of both joints of each of the five fingers.

One of the major disadvantages of previous approaches to multi modal interaction with embodied conversational agents is the inability of agents to respond to user actions before those actions are completed, thus giving rise to the delay in the reaction of the agent and making the dialogue not natural. For example, even a simple gazing at the user when the user starts the movement can make the conversation more natural.

In our work we made an attempt to solve this problem by recognizing the user's behavior at an early stage.

As mentioned above, we use a set of sensors to detect movements of the user. However, this equipment gives us the only information on the changing of coordinates of an inertial sensor in space. We use the Self-Organizational Incremental Neural Network (HB-SOINN) (Okada, Nishida, 2009) for gestures recognition.

Once the user starts to perform an action, a probability ranking of trained gestures is created and forwarded to the Fusion component where a ranking of possible user intended behaviors is created and forwarded to the Human Behavior Interpreter which interprets the possible behavior according to the current context of discourse. Then the multi-modal fusion component collects information on the four user modalities and merges them into the single joint output.

To achieve a more natural interaction between user and agent it is not necessary to enforce a strict timeline on user gestures forming an utterance. In a conversation people sometimes use hand gestures before they complement it with speech and vice versa. A parsing of multiple modalities which requests the different modalities to be taken out in predefined temporal alignment such as a state device would therefore result in a lower recognition rate of coherent gestures and lead to an unnatural way of interaction with the virtual human. Moreover it is not always necessary to take all four modalities into account to form a joint interpretation. Sometimes a spoken word is enough to express an intention and additional gestures do not enhance the content of information. However, some situations require for additional gestures such as pointing out to a direction or describing a path.

Coherent modalities forming user behavior are modeled allowing for the different modalities to be recognized independently from each other by representing them as units according to their necessity of coherence. For example the utterance of pointing out an object on the screen and additionally expressing which type of object the user is pointing to by speech consists of three modality types (hand, arm and speech). A unit can be imagined as a complete graph with input types as vertices and transitions as events when a new modality is processed. A user behavior is successfully recognized if all nodes of the graph are visited.

6.1.2 Human Behavior Interpretation (HBI Component)

The natural interaction between the agent and the user requires an adequate agent's response to the actions of the user in the context of the current situation. In various situations the same action of the the user can have different meanings. For example, raised up the hand in the beginning of the conversation may be interpreted as a greeting, while during the conversation it can mean the desire to interrupt the agent or to attract attention. To resolve such situations, we use the component of the interpretation of the user behavior (Human Behavior Interpreter, HBI). This component receives information about user actions from the Human Behavior Detector, compares user actions with the current situation and interprets the user actions as intentions.

According to the principles of its work, the component directly interacts with the Context.

6.1.3 Context

The Context stores all the relevant information for initialization and processing of our agent. Context describes the object, its size and location, the meta-information about the environment. The coordinates of the object plays an essential role in the correct interpretation of user's gestures, such as when pointing to any portion of the facility. Meta-information is used to describe the object's relationship with the environment (above, below, before, for), and visual characteristics of the object(color, shape). In other words, the meta-information is used to identify the objects using verbal descriptions.

In addition, the context stores such personal and interpersonal data as names, gender, age, language, relationships. This information is needed to properly interpret and generate the intentions and behaviors.

6.1.4 Intention Planning

We use Jadex Framework (Braubach, Pokahr, Lamersdorf, 2004) for intentions planning, which implements the Belief-Desire-Intention (BDI) model of decision-making.

Belief-Desire-Intention model has been proposed by Bratman as a theory of human practical reasoning (Bratman, 1987). The principles of the model are based on simplifying complex procedures and dependencies in the decision-making to the motivation stance. The actions of the agent based only on his desires, while ignoring other possible causes. Thus, BDI model implements a clearly rational planning.

Jadex Framework supports the creation of our own beliefs, desires and intentions for the Virtual Agent. During runtime the agent is able to load his beliefs out of the context and can update them after processing an input. Architecture of Intention Planner is shown on fig 2.

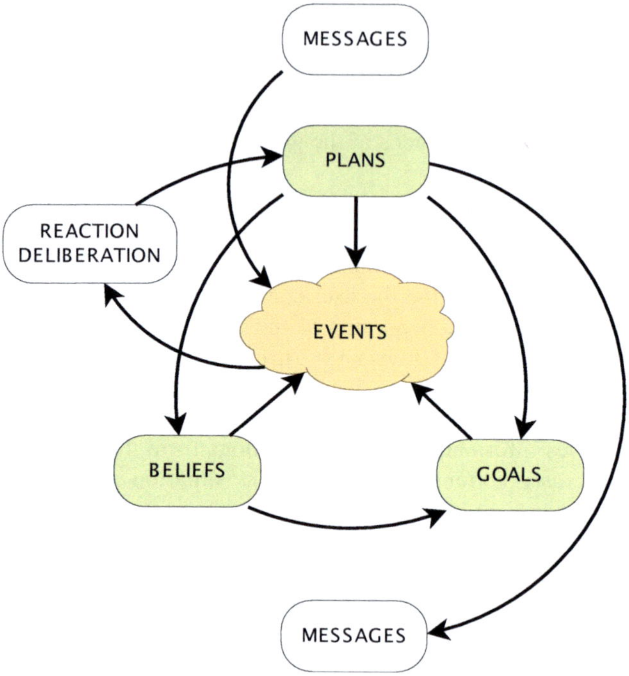

Fig. 2. Architecture of the Intention Planner

Events are the basis of this component. The input information for the Intention Planner are the messages that generate new events. Input events in conjunction with the internal events of the Planner are input to the internal reaction and deliberation mechanism, which matches plans from the library to these events. In Jadex architecture only deliberation mechanism is global. All other components are grouped into the reusable modules.

To generate intentions Intention Planner receives a messages from the Human Behavior Interpreter, and uses the Context. In this architecture Intention Planner also receives reports from the Emotion Engine, that can be processed in combination with the events of the Human Behavior Interpreter.

In the output of the Intent Planning stage the intentions are described by using our own attempt of realizing a Function Markup Language called xGFML. Our version of xGFML supports descriptions for personal information about the agent and the user, interpersonal information, a mental state and information concerning the conversational act.

6.1.5 Behavior Planning

The Behavior Planning generates GSML described behaviors out of the intentions. Intentions described in xGFML provide input for the Behavior Planning. By using different dictionaries the communicative-tag of the xGFML description is substituted. The dictionaries are chosen on base of the language-attribute. Within the dictionaries are building blocks for behaviors including vocal output und animations. The combinations of these behaviors differ in the dictionaries depending on the language and associated culture. Therefore a "greet" may be substituted by "Hello..." and waving hand for English language in contrast to "Konnichiwa…" and taking a bow for Japanese language. Additional information within the xGFML description (e.g. emotion, personal and interpersonal information) are used for modifying the way the behavior are realized (e.g. speed and pitch of verbal output, smiling or serious facial expressions).

6.1.6 Behavior Realization

The stage of Behavior Realization is implemented with the VisagePlayerM. VisagePlayerM is an extended version of the Visage Player of Visage Technologies for the Generic ECA framework. It uses GECA's markup language GSML for describing the agent's behavior.

6.2 Integrating Processing of Emotional Information

Surely, some primitive functions of pseudo-emotional behavior can be integrated as a functions of the pure reasoning agent. But this way seems to be a bit dangerous due to several reasons, the main of which is functional restrictions. For instance, this way allows to implement perception and expressing of emotions and a kind of pure reasoning about emotional information within the BDI core while loosing functions related to influencing cognitive functions (for example, goal deliberation) by emotions.

We developed an extended architecture for processing emotional information having in mind previously mentioned appraisal theory of emotions and theory about primary and secondary emotions. Additionally we distinguish between three kinds of emotional information which differ in their nature and essence.

First of all, the agent is expected to "have" emotions. Emotions are immediate reactions to external or internal events; they can have latencies from seconds to minutes and they generate noticeable behavior.

Emotions also can influence changes on the deeper layer which is the mental state of the agent. Obviously, mental state can not be changed so easily as transient emotions. On the other hand, mental state influences emotion generation. For example, being in a bad mood one can react not so intensively on a positive event, than one who is in good mood. Latencies of mental states can vary from hours to days. On the other hand, mental states have weaker behavior generation capabilities than emotions.

The third and deepest layer is a temperament or inclinations (preferable term) which is constant (infinitive latency) and defines a point of stability when system does not have any events and inclinations in changing a mental state in response to emotions. Inclinations do not have any explicit influences over behavior.

An advantage of this approach is that this three-layered model allows us to implement parallel and different processing of different kinds of emotional information in the structured model, refraining from the joint processing of a some kind of medley of emotional information.

Let us recall last strings of the third paragraph of this chapter. We concluded that emotions are parameters of behavior and among others we can observe emotions by observing behavior. Noticeable that behavior can not be observed from image; we need a some kind of short film to understand one's behavior and, respectively, emotions. For instance, while looking at the smiling face on a picture we can guess that person is satisfied or happy, but by observing a short film (behavior) we can understand that person shows an irony and his/her mental state is not so positive.

The same idea is applied for expressing behavior by the agent. The architecture, based on this three-layered model and which realizes the appraisal and primary-secondary theories of emotions is shown in fig. 3.

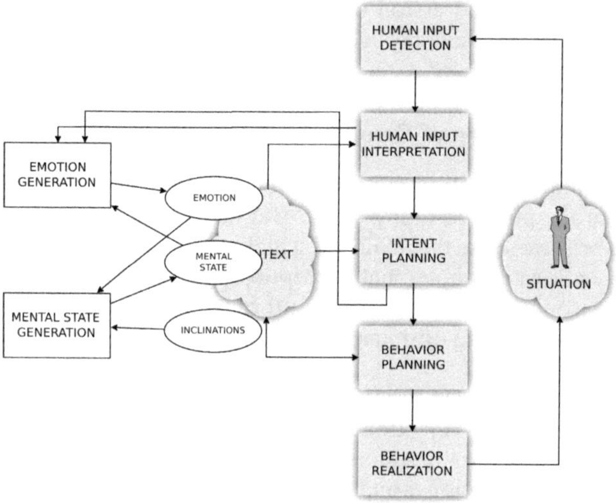

Fig. 3. Components for processing emotional information

Since emotions are appraisals of events, emotion generation block receives the same messages as the intent planner from the human input interpretor. Thus we can perform a concurrent pure-reasoning and emotional processing of the same input information. Current emotion is stored into the context and is used in behavior planning stage. This is a path of primary emotions. Secondary emotion path works when intent planner sends a query to generate a particular emotion to emotion generator. This secondary emotion is also stored in the context and used by behavior planner to chose particular actions and parameters of actions (for example, intensity). When generating a current primary emotion emotion generator reads a current mental state from the context. At the same time the mental state generator continuously reads a current emotion and changes current mental state according to this emotion and inclinations. Both current emotion and current mental state are be used by the intention planner.

In our current model we employ PAD emotion scales for describing emotion and mental state of the agent. We supposed that the agent has no inclination and when no events happens agent gets neutral.

6.3 Scenario

For our project we developed a complete scenario in which agent plays different roles depending on the state of the device. Here we would like to show a simple scene which is relevant for this chapter. In this scene agent play the role of the Virtual Technical Support Engineer trying to figure out the problem of malfunctioning of the device.

Scene: device's malfunctioning
User says that device does not work. This is unexpected event and agent's primary emotion is surprise. After that agent becomes serious and concentrated showing it's readiness to help using the secondary emotion path. Then agent tries to help and when problem is figured out agent realizes primary emotion according to goal achievement event. After that agent again shows it's readiness to solve a problem, but emotion is tinctured by agent's positive mental state.

Messages between modules together with comments can be found in tab. 1.

Table 1. Messages for scenario

Component	Action	Comments
HBD	**keyword: doesn't, work gesture: none**	*User says that device does not work.*
HBI	**event(unexp, 'device', malfunctioning);**	*New event about device's malfunctioning generated and sent to EG and IP*
EG	**primary('surprised');**	*This event is unexpected and agent gets surprised*
BP	**stopCurrentAction(); outEmotion();**	*Behavior Planner interrupts current action and exhibits primary emotion*

IP	**currentPlan.stop();
	saveState(currentPlan);
	findProblem();
	setEmotion('concentrated');**
		Agent stops a current plan and generates behavior which shows it readiness to help

EG	**secondary('concentrated');**	*Secondary emotion generated*
BP	**outEmotion();**	*Secondary emotion realized*
....		*Agent asks questions trying to find out the problem; final question is "what is you paper size?"*
HBD	**keyword: bee, five gesture: none**	*User answers that paper size is B5*
HBI	**Event(answer, 'paper', B5);**	*But printer can use only A4 paper size*
IP	**setEmotion('satisfied');**	*Agent shows that he understood the problem by expressing satisfaction*
EG	**primary('satisfaction');**	
BP	**outEmotion();**	
IP	**inform('device', 'paper size');**	*After that agent starts explanation about usable paper size*
BP	**queryDecription('device', 'paper size');**	

7 Evaluation Methodology

Any changes in the behavior of the agent leads ultimately to a change in the quality of interaction between human and the virtual agent. But how can we ascertain that the changes actually improve the quality of interaction, rather than vice versa?

To solve this problem, we compare two different versions of virtual agents with each other, asking users to answer a questionnaire and pass a test (these methods are discussed in details below). As a rule, we compare the new version of the agent, which had been amended with the previous one. Thus, we can observe the trend of improvement or worsening the quality of agent's interactions with humans.

This method has several significant shortcomings. Firstly, there should be a large number of subjects in order to obtain reliable results because this is a statistical approach. Secondly, it is necessary to keep a balance between the significant and minor modifications due to the following reasons: minor changes (and consequently a small difference in the behavior) of the two agents may simply not affect the results of the experiment, but on the other hand, when making a large number of changes, it may be simply impossible to say which of the imposed innovation has improved (or worsened) the quality of the agent.

In addition, when comparing the three agents, of which, for example, the second agent is better than the first one, and third agent is worst than second one, no one can say that the first agent is better or worse than the third agent, because we do not have any quantitative measure.

Despite this, this approach is one of the main tools of experimentation due to its relative simplicity and versatility, not to mention the fact that we simple do not have a universal quantification, which allows agents to evaluate the absolute scale. The following sections consider in more detail methods we use.

7.1 Questionnaire

Questionnaire is primarily a statistical tool to identify an attitude of the group of subjects to objects. In our case objects are virtual agents.

Pioneer in using of questionnaires in a psychological study is Sir Francis Galton, who in his study on the influence of heredity and environment on the level of intellectual achievement with the help of the questionnaire polled a hundred of outstanding British scientists.

Questionnaires have advantages over some other types of surveys in that they are cheap, do not require as much effort from the questioner as verbal survey, and often have standardized answers that make it simple to compile data. However, such standardized answers may frustrate users.

In our work we use mostly questionnaires with scaled questions. For instance we ask subjects to evaluate how smooth was conversation with the agent by using scores from 1 (not at all) to 5 (very smooth). This allow us to obtain precision enough statistical data even from the small group of subjects.

7.2 The Implicit Association Test

Another method we use in experiments is the Implicit Association Test (IAT).

The Implicit Association Test (IAT) (Greenwald, McGhee, Schwartz 1998) is an experimental method within social psychology designed to measure the strength of automatic association between mental representations of objects (concepts) in memory. The IAT requires the rapid categorization of various stimulus objects, such that easier pairings (and faster responses) are interpreted as being more strongly associated in memory than more difficult pairings (slower responses).

The IAT was introduced in 1998 by Anthony Greenwald, Debbie McGhee, and Jordan Schwartz. Project Implicit, a research and educational outreach program that allows individuals to take the test over the web, is run by social psychologists Greenwald, Mahzarin Banaji, and Brian Nosek.

The IAT allows us to obtain implicit users' preferences concerning a particular agent by comparing this agent with another one (usually with older version of the agent or the agent with disabled key features). This assessment is independent on possible users' intentions to show particular results.

As we see from the classic procedure and scoring algorithm the test is designed for evaluating attitudes towards the well-known concepts. For instance, one of classic examples of this test allows to evaluate subject's attitude towards flowers and insects by showing names of flowers and insects.

7.3 Experiment Design

One of the most important issues of evaluating a new agent is an experiment design. Actually, we would say that there are no non-significant things in the procedures of experiment. First of all, we make subjects try to take the agent as a real person. We also ask them to communicate with the agent in the natural way as they communicate with other people.

Experiment consists of two parts: conversation with agents and evaluation. At the first part of experiment we ask participants to interact with agents For example, we ask users to learn as much as possible about two slightly different target devices by communicating with agents which can differ by functionality. Target devices can different logo, power class, color, slightly different functionality. Each participant has to communicate sequentially with two agents, which explain it's device.

Participants are not limited in time; they can communicate with agents as much as they want asking such question as "what is this button for (along with pointing gesture)?" or "which type of paper can I use?". Due to technological limitation agents can not recognize full meaning of user's speech and comprehensive gestures, therefore we give to users some kind of dictionary containing words and gestures, suggesting but not limiting subjects to use this dictionary. Before experiment we explain the goal and rules of experiment and during experiment we refrain from any disturbing or interrupting the experiment even caused by misunderstanding between subjects and agents. After communicating with each of the agents we ask users to fill out the questionnaire which contains questions about devices which subjects are expected to learn and their feelings about conversation.

After interacting with two agents and filling out questionnaires we ask subjects to take the implicit association test, in which is devoted to evaluate their attitudes towards two different devices. Thus we compare agents indirectly, by comparing two devices presented by agents.

This approach seems to be a bit incorrect due to the fact that the IAT is devoted for evaluating attitudes towards well-known concepts, as we mentioned above. This means that in classic IAT application all mistakes are caused by misprinting, but in our experiment there can be misprints and also mistakes. This can cause a noise and we have to eliminate results with a high portion of mistakes. Though, this approach gives enough precision results if difference between agents' functionality is noticeable.

8 Conclusions

This chapter addresses the problem of building the Virtual Technical Support Engineer which plays a role of a kind of media supporting user to figure out the problem, delivering relevant information (solution) and getting feedback from user which can help to evaluate usability and effectiveness of the agent. In this chapter we show what is an emotional intelligence as a part of social (interpersonal) intelligence and why is it important to integrate emotional intelligence into virtual humans. We use scenario to show how integrating the processing of emotional information into the pure-reasoning agent can bias agent's decision making and behavior and how these features can be used to achieve a natural face-to-face communication between agents and humans.

This chapter shows some theoretical background about features of emotions supposing that emotions can be defined as parameters of behavior. We also show three computational models of emotions which can be used when building virtual agents.

This chapter contains the evaluation methodology which we use in out work along with an experiment design notes, which are very important to build a high-quality experiment.

In our work we do not pay attention to the pure reflex behavior, which is also important to achieve natural communication. Realization of these functions in the dedicated module (as, by the way, they have been already realized in living organisms) will relieve the intent planner (where they are currently realized) for other functions and increase a speed of agent's reaction.

Evaluation methodology also should be modified. The first problem is that there is a difference between classic and our usage of the IAT. We use it for not-well-known categories, which causes mistakes (in addition to misprints). This can cause inaccurate results and we have to eliminate results with a high portion (>25%) of mistakes instead of effective using of this information.

Secondary, the combination of questionnaires with the IAT does not give accurate enough results and other additional methods should be used. It can be physiological measurements or/and facial expression analysis which can give immediate results in the contrast to IAT and questionnaire which give results only after experiment.

References

Ascension Technology Corporation, `http://www.ascension-tech.com`

Boyatzis, R., Goleman, D., Rhee, K.: Clustering competence in emotional intelligence: insights from the emotional competence inventory (ECI). In: Bar-On, R., Parker, J.D.A. (eds.) Handbook of emotional intelligence, pp. 343–362. Jossey-Bass, San Francisco (2000); Cassell, J., Sullivan, S.: Embodied Conversational Agents. MIT Press, Cambridge (2000)

Bradberry, T., Greaves, J.: The Emotional Intelligence Quick Book. Simon and Schuster, New York (2005) (ISBN 0743273265)

Bratman, M.: Intention, Plans, and Practical Reason. Harvard University Press, Cambridge (1987)

Braubach, L., Pokahr, A., Lamersdorf, W.: Jadex: A Short Overview. In: Net.Object Days 2004: Agent Expo. (2004)

Cavallo, K., Brienza, D. (2002), Emotional competence and leadership excellence at Johnson and Johnson: the emotional intelligence leadership study. Website: `http://www.eiconsortium.org`

Damasio, A.: Descartes' Error: Emotion, Reason, and the Human Brain. Putnam Publishing, New York (1994)

Gardner, H.: Frames of Mind: The Theory of Multiple Intelligences. Basic Books, New York (1983)

Gottfredson, L.: Mainstream Science on Intelligence. Wall Street Journal A18 (December 13 1994)

Greenwald, A.G., McGhee, D.E., Schwartz, J.K.L.: Measuring individual differences in implicit cognition: The Implicit Association Test. Journal of Personality and Social Psychology 74, 1464–1480 (1998)

Herrnstein, R.J.: Murray Charles, The Bell Curve (1994) (ISBN 0-02-914673-9)

Huang, H., Cerekovic, A., Pandzic, I., Nakano, Y., Nishida, T.: The Design of a Generic Framework for Integrating ECA Components. In: Proceedings of 7th International Conference of Autonomous Agents and Multiagent Systems (AAMAS 2008), Estoril, Portugal, pp. 128–135 (May 2008)

Leuner, B.: Emotional intelligence and emancipation. Praxis der Kinderpsychologie und Kinderpsychiatrie 15, 193–203 (1966)

Loquendo Vocal Technology and Services, `http://www.loquendo.com`

Mayer, J.D., DiPaolo, M.T., Salovey, P.: Perceiving affective content in ambiguous visual stimuli: a component of emotional intelligence. Journal of Personality Assessment 54, 772–781 (1990)

Mehrabian, A.: Framework for a comprehensive description and measurement of emotional states. Genetic, Social, and General Psychology Monographs 121, 339–361 (1995)

Mindmakers.org SAIBA Multimodal Behavior Generation Framework, `http://wiki.mindmakers.org/projects:SAIBA:main`

Neisser, U., Boodoo, G., Bouchard Jr., T.J., Boykin, A.W., Brody, N., Ceci, S.J., Halpern, D.F., Loehlin, J.C., Perloff, R., Sternberg, R.J., et al.: Intelligence: Knowns and Unknowns. In: Annual Progress in Child Psychiatry and Child Development 1997 (1998)

Okada, S., Nishida, T.: Online Clustering of Gesture Patterns based on Self-organizing Incremental Neural Network. In: IEEE International Joint Conference on Neural Networks, IJCNN 2009 (2009)

Ortony, A., Clore, G.L., Collins, A.: The Cognitive Structure of Emotions. Cambridge University Press, Cambridge (1988)

Payne, W.L.: A study of emotion: developing emotional intelligence; self integration; relating to fear, pain and desire. Dissertation Abstracts International 47, 203A (1983) (University microfilms No. AAC 8605928)

Plutchik, R., Conte, H.R.: Circumplex Models of Personality and Emotions. American Psychological Association, Washington (1997)

Salovey, P., Grewal, D.: The Science of Emotional Intelligence. Current directions in psychological science 14(6) (2005)

Salovey, P., Mayer, J.D.: Emotional intelligence. Imagination, Cognition, and Personality 9, 185–211 (1990)

Scherer, K.R., Shorr, A., Johnstone, T. (eds.): Appraisal processes in emotion: theory, methods, research. Oxford University Press, Canary (2001)

Thorndike, E.L.: Intelligence and its use. Harper's Magazine 140, 227–235 (1920)

Appendix

Appendix A: Complete Target Scenario

In this complete scenario we made an attempt to reproduce all important stages of a novel device. The first and one of the most important steps after developing (or at the same time with developing) is to show this device to potential customers. It may be, for example, showing device on a specialized exhibition. Some possible scenes of this stages are reproduced in the first part of the complete scenario. Here agent gives a basic introduction and user can ask agent about features of the device. In this part of the scenario there is no long-term relationships between users and the agent.

After that, customer might be interested in some getting started guide. These scenes can be found in the Part 2. This part is almost the same as part 1 excepting the fact that there is a possibility for long-term relationships between customer and agent.

Part 3 addresses situations with device's malfunctioning. One scene from this part we use in this chapter to show abilities of the agent for emotional information processing.

Part I: Device Presentation

Scene 1.1: Greeting
Agent is presenting himself and asks user about his/her name in the final stage of this scene. Agent shows positive emotions.

Scene 1.2: Initial Stage
After the engineer finished with greeting and introducing himself, he gives the customer the possibility to inform him about his interests. If there is no special interest the engineer starts with giving general information about the device to the customer and gets more detailed in the course of the conversation. In this scene the virtual engineer gets no information and starts with his initial plan (no input behavior).

Scene 1.3: Explicit Interruption
During this description the customer is interrupting the engineer and asking information about a button of the device and pointing at the device ("What is this button for?") The engineer will request confirmation about the part of the device and then starts to explain. Finishing the explanation concerning the button without getting further advices, the engineer will continue the general explanation of scene 1. In this scene the engineer reacts on a direct request of the customer including a pointing gesture and a spoken question (active input behavior).

Scene 1.4: First distraction
During the continuation of the general description the customer shows sign of distraction. He is turning away from the scene and focusing a different part of the

device. The engineer is taking notice of the distraction, but continues his explanation. The engineer believes that the single sign of distract is not sufficient for. The customer showed a passive input behavior that is not significant enough to trigger a change in the plan at the moment.

Scene 1.5: Second distraction, explicit topic selection
Again the customer is distracted by particular part of the device. Since it was not the first sign of distraction the engineer believes that the customer is more interested in information about the object he's distracted by than about a general explanation. So, he stops interrupts the current topic, confirms the new topic and begins to talk about the object of distraction. In this scene the engineer reacts on a passive behavior combined with a part of the device.

Scene 1.6: Third distraction
During the continuation the customer is distracted again. This Scene is similar to scene 3 with the difference that the reason for the distraction is not known. The engineer takes notice, but continues his explanation.

Scene 1.7: User's disinterest
The customer is still/again distracted without showing interest to other parts of the device. The engineer estimates this as a sign of disinterest for the current conversation. He stops the explanation and asks for other interest of the customer without a result. So he has to change the topic based on his own beliefs and without an input by the customer (only that the current topic was not interesting).

Part II: Getting Started

Part III: Malfunctioning

Complex scene: figuring out the problem
User says that device does not work. This is unexpected event and agent's primary emotion is surprise. After that agent becomes serious and concentrated showing it's readiness to help using the secondary emotion path. Then agent tries to help and when problem is figured out agent realizes primary emotion according to goal achievement event. After that agent again shows it's readiness to solve a problem, but emotion is tinctured by agent's positive mental state.

Chapter 9
Recognition of Fine-Grained Emotions from Text: An Approach Based on the Compositionality Principle

Alena Neviarouskaya[1], Helmut Prendinger[2], and Mitsuru Ishizuka[1]

[1] Department of Information and Communication Engineering, University of Tokyo, R. 111C1/111D2, Eng. Bldg. 2, 11th floor, 7-3-1 Hongo, Bunkyo-ku, Tokyo 113-8656, Japan
lena@mi.ci.i.u-tokyo.ac.jp, ishizuka@i.u-tokyo.ac.jp
[2] Digital Content and Media Sciences Research Division, National Institute of Informatics 1613-1B, 16 floor, 2-1-2 Hitotsubashi, Chiyoda-ku, Tokyo 101-8430, Japan
helmut@nii.ac.jp

Abstract. This chapter addresses the tasks of recognition, interpretation and visualization of affect communicated through text messaging in virtual communication environments. In order to facilitate sensitive and expressive communication in such environments, we introduced a novel syntactic rule-based approach to affect recognition from text. Our Affect Analysis Model follows the compositionality principle, according to which emotional meaning of a sentence is determined by composing parts that correspond to lexical units or other linguistic constituent types governed by the rules of aggregation, propagation, domination, neutralization, and intensification, at various grammatical levels. The proposed rule-based approach processes each sentence in sequential stages, including symbolic cue processing, detection and transformation of abbreviations, sentence parsing, and word/phrase/sentence-level analyses. Our method is capable of processing sentences of different complexity, including simple, compound, complex (with complement and relative clauses), and complex-compound sentences. Affect in text is classified into nine emotion categories (or neutral), and, additionally, information that indicates social communicative behaviour is identified. The evaluation of the Affect Analysis Model algorithm showed promising results regarding its capability to accurately recognize affective information in text from an existing corpus of informal online conversations. The applications of the developed Affect Analysis Model in Instant Messaging system (AffectIM) and in Second Life (EmoHeart, iFeel_IM!) are described in the chapter.

1 Introduction and Motivation

'Emotion is what gives communication life. A conversation between emotionally involved partners is bright and lively, but a meeting without feeling is deadly dull.' (Planalp 1999)

T. Nishida et al. (Eds.): Modeling Machine Emotions for Realizing Intelligence, SIST 1, pp. 179–207.
springerlink.com

Emotions play the role of a sensitive catalyst, which fosters lively interactions between human beings and assists in the development and regulation of interpersonal relationships. The expression of emotions shapes social interactions by providing observers a rich channel of information about the conversation partner (Ekman 1993) or his social intentions (Fridlund 1992), by evoking positive or negative responses in others (Dimberg and Ohman 1996), and by stimulating other's social behaviour. Keltner et al. (2003) highlight that 'facial expressions are more than just markers of internal states', which also serve unique functions in a social environment. By accentuating the functional role of emotions, Frijda (1986, 1994) argues that they preserve and enhance life, and Lutz (1988) emphasizes their communicative, moral, and cultural purposes.

The richness of emotional communication greatly benefits from the expressiveness of verbal (spoken words, prosody) and nonverbal (gaze, face, gestures, body pose) cues that enable auditory and visual channels of communication (Planalp 1999). All types of expressive means potentially carry communicative power and promote better understanding (Allwood 2002). Emotional information can be (1) encoded *lexically* within the actual words (affective predicates, intensifiers, modals, hedges, etc.) of the sentence, *syntactically* by means of subordinate clauses, and *morphologically* through changes in attitudinal shades of word meaning using suffixes (especially, in languages with rich inflectional system, such as Russian or Italian); (2) consciously or unconsciously conveyed through vast repertoire of prosodic features like intonation, stress, pitch, loudness, juncture, and rate of speech, (3) visually reflected through subtle details of contraction of facial muscles, tints of facial skin, eye movements, gestures, and body postures (Reilly and Seibert 2003). The emotional significance of an utterance is accompanied, complemented and modified by vocal and visual cues.

Nowadays, media for remote online communications and emerging 3D virtual worlds that provide new opportunities for social contact grow rapidly, engage people, and gain great popularity among them. The main motivations for 'residents' of chat rooms or virtual environments to connect to these media are seeking conversation, experimenting with a new communication media, and initiating relationships with other people. A study conducted by Peris et al. (2002) revealed that 'relationships developed online are healthy' and considered by people 'as real as face-to-face relationships'. Findings described in (Hu et al. 2004) indicate that there is a positive relationship between the amount of online media use and verbal, affective, and social intimacy, and that frequent online conversation actually encourages the desire to meet face-to-face, reinforcing thus personal interaction. To emphasize the realism and significance of social exchanges in such environments, Chayko (2002) recently proposed to use the term 'sociomental' rather than 'virtual'. Without doubt, computer-mediated communications that facilitate contact with others have strong potential to affect the nature of social life in terms of both interpersonal relationships and the character of community (Rheingold 2002; Fernback 2004).

To establish a social and friendly atmosphere, people should be able to express emotions. However, media for online communication lack the physical contact and the visualization of emotional reactions of partners involved in a remote text-mediated conversation, thus limiting the source of information to text messages and to graphical representations of users (avatars) that are to some degree controlled by a person. Trends show that people often try to enrich their interaction online, introducing affective symbolic conventions or emphases into text (emoticons, capital letters etc.) (Reid 1991; Walther and D'Addario 2001; Hu et al. 2004; Yigit 2005; Derks 2007), colouring emotional messages, or manually controlling the expressiveness of avatars in order to supplement the lack of paralinguistic cues. One of the first attempts to study effects of conveying emotional expressions through communication in computer-mediated environment was done by Rivera et al. (1996). The results of their experiment indicated that subjects allowed to use emoticons were more satisfied with the system than those subjects having conversations without these symbolic emotional expressions. The user study of ExMS (Persson 2003), a messaging system that allows its users to concatenate and annotate avatar animations, showed that interplay between pure text and animation significantly improved the expressiveness of messages, and that users felt pride of being identified with their embodied representation.

The motivation behind our research on affect sensing from text is to enrich social interactivity and emotional expressiveness of real-time messaging, where a machine is used as a communication channel connecting people and transmitting human emotions. Here, a key issue is to provide the automation of multiple expressive means so that the user does not have to worry about visual self-presentation as in standard Instant Messaging (IM) systems, but can focus on the textual content of the conversation.

In this research, we address the tasks of recognition and interpretation of fine-grained emotions communicated through text messaging in virtual communication environments, where people tend to use an informal and abbreviated style of writing, often accompanied by the use of emoticons (short symbols that resemble facial displays). The evolving nature of language in online conversations is a main issue in affect sensing from this media type, since sentence parsing might fail in syntactic structure analysis. In order to facilitate sensitive and expressive communication in computer-mediated environments, we introduced a novel syntactic rule-based approach to affect recognition from text. The key feature of our approach is that it is designed to handle not only grammatically and syntactically correct textual input, but also informal messages written in an abbreviated or expressive manner (Neviarouskaya et al. 2007a). In contrast to machine-learning based approaches, for which detection of fine-grained emotions on a sentence level still remains a challenge, the proposed rule-based approach processes each sentence in sequential stages, including symbolic cue processing, detection and transformation of abbreviations, sentence parsing, and word/phrase/sentence-level analyses. Thus, each analyzed sentence is automatically annotated with an emotion (or neutral) label and a numerical value indicating the strength of emotion.

The remainder of the chapter is structured as follows. In Section 2, we report on related work. Section 3 summarizes the algorithm for recognition of fine-grained emotions from text, and describes the experimental results. The applications of the developed Affect Analysis Model in Instant Messaging system (AffectIM) and in

Second Life (EmoHeart, iFeel_IM!) are described in Section 4, Section 5, and Section 6, respectively. Finally, Section 7 concludes the chapter.

2 Related Work

The emergence of the field of affective computing (Picard 1997) has greatly inspired research challenging the issues of recognition, interpretation and representation of affect. The emotional information expressed through a wide range of modalities has been considered, including affect in written language, speech, facial display, posture, and physiological activity. According to Picard (1997),

> 'the basic requirement for a computer to have the ability to express emotions is that the machine has channels of communication such as voice or image, and an ability to communicate affective information over those channels...'.

Physiological bio-signals (such as facial electromyogram, electrocardiogram, respiration effort, and electrodermal activity) were analysed by Rigas et al. (2007) to define emotional state of a human. Recent studies have begun to illuminate how emotions conveyed through vocal channel can be detected (Slot et al. 2008; Wu et al. 2009). Visual information also carries valuable emotional content. Considering the fact that eye and mouth expressions are most evident emotional indicators on the face, Maglogiannis et al. (2009) developed a method that, based on color images, detects skin, eye and mouth regions, and recognizes emotions encoded in these clues by detecting edges and evaluating the color gradient. Aimed at the synchronization of the avatar's state in virtual environment with the actual emotional state of the user, Di Fiore et al. (2008) realized the automatic extraction of emotion-related metadata (particularly, facial features) from a real-time video stream originating from a webcam. Castellano et al. (2008) have proposed a multimodal approach for emotion recognition from facial expressions, body movement, gestures, and speech. After training individual Bayesian classifiers for each modality, researchers fused the data at both feature and decision levels, that resulted in the increase of accuracy compared to the unimodal approach.

The most challenging tasks for computational linguists are classification of text as of subjective or of factual nature, determination of orientation and strength of sentiment, and recognition of attitude type expressed in text. Various approaches to subjectivity, sentiment or affect analysis on different textual composition levels were proposed: (1) word level (Subasic and Huettner 2001; Kamps and Marx 2002; Turney and Littman 2003; Baroni and Vegnaduzzo 2004; Andreevskaia and Bergler 2006; Strapparava et al. 2007); (2) synset level (Esuli and Sebastiani 2006); (3) phrase level (Wilson et al. 2005); (4) clause or sentence level (Olveres et al. 1998; Boucouvalas 2003; Liu et al. 2003; Mulder et al. 2004; Read 2004; Kim and Hovy 2005; Neviarouskaya et al. 2007b; Moilanen and Pulman 2007); (5) paragraph or document level (Turney 2002; Pang et al. 2002; Mishne 2005; Kim and Hovy 2006; Leshed and Kaye 2006; Mihalcea and Liu 2006; Nadeau et al. 2006).

WordNet-Affect, a linguistic resource for the lexical representation of affective knowledge, was created by Strapparava and Valitutti (2004). Automatic textual emotion recognition and its visualization by kinetic typography (text animation) are described in (Strapparava et al. 2007). In order to analyse affective content, the authors were using not only affective words from WordNet-Affect, but also an indirect affective lexicon derived from the evaluation of the semantic similarity between generic terms and affective lexical concepts using Latent Semantic Analysis.

Kamps and Marx (2002) investigated measures for affective and emotive aspects of meaning obtained from the structure of the WordNet lexical database (Miller 1990). Since the meaning of a concept in WordNet is determined by its position relative to other concepts, the authors decided to evaluate individual words (specifically, adjectives) by determining their relation (or distance) to the words 'good' and 'bad', and assigning values in the interval [-1;1]. As to evaluation of larger textual units, a straightforward aggregation procedure was implemented, where text was represented as a bag of words, and the scores of individual words were simply added up. An approach to analysing affect content in free text using fuzzy logic techniques was proposed by Subasic and Huettner (2001). Research on mining WordNet for fuzzy sentiment was conducted by Andreevskaia and Bergler (2006).

To classify sentiment and affect represented in text, methods employing Point-wise-Mutual Information calculation were introduced (Turney 2002; Read 2004). Turney (2002) presented an unsupervised learning algorithm for classifying reviews as 'recommended' or 'not recommended' based on the estimation of the average of the semantic orientation of phrases. In contrast to Turney's (2002) work, Pang et al. (2002) applied several completely prior-knowledge-free supervised machine learning methods, with the goal of classifying film reviews into 'positive' and 'negative' ones. Some researchers (Turney 2002; Pang et al. 2002) observed that for more accurate classification some form of discourse analysis is necessary, or at least some way of determining the focus of each sentence as, according to Turney (2002: 424), 'the whole is not necessarily the sum of the parts'.

Wilson et al. (2005) presented experiments in which they automatically distinguish prior and contextual polarity of individual words and phrases in sentiment expressions, in contrast to classifying documents by their overall sentiment. Kim and Hovy (2005) developed an automatic algorithm for classifying opinion-bearing and non-opinion-bearing words, and described a method for the detection of sentence-level opinion. They built a classifier that identifies all sentences expressing valence in a given text, using strong markers of opinion such as certain modal verbs, adjectives and adverbs. Unsupervised statistical method for the task of separating opinions from facts using Naïve Bayes classifier, proposed by Yu and Hatzivassiloglou (2003), resulted in high accuracy (up to 91 %) at sentence level. The best performance was observed when words, bigrams, trigrams, part-of-speech, and polarity were included in the feature set. The decision on the polarity of an opinion sentence was based on the number and strength of semantically oriented words in the sentence.

In order to overcome the problem of strong dependency of machine learning techniques on domain, topic and time, Read (2005) constructed a corpus of text marked-up with emoticons and developed the emoticon-trained classifier aimed at sentiment classification. While this classifier performed well (up to 70 % accuracy) on the articles extracted from the constructed corpus, it was not very effective in predicting the valence of movie reviews and news. Read (2005) speculated that there is a language-style dependency in sentiment classification. Researchers also applied statistical language modelling techniques to analyse moods conveyed through online diary-like posts (Mishne 2005; Leshed and Kaye 2006; Mihalcea and Liu 2006). However, machine learning approaches to affect classification in text suffer from limitations such as: necessity of large amounts of data to gather meaningful statistics; neglect of negation and condition constructions; and disregard of syntactic relations and semantic dependencies in sentences.

Some researchers employed a keyword-spotting technique to recognize emotion in text (Olveres et al. 1998; Strapparava et al. 2007). However, the use of a purely word-level analysis model cannot handle cases where affect is expressed by phrases requiring complex phrase/sentence-level analyses, as words are interrelated and influence each other's affect-related interpretation (as in the sentence '*I use the ability to breathe without guilt or worry*'), or when a sentence carries affect indirectly through underlying meaning (for example, '*I punched my car radio, and my knuckle is now bleeding*').

Advanced approaches targeting textual affect recognition at the sentence level are described in (Boucouvalas 2003; Liu et al. 2003; Mulder et al. 2004; Alm 2008). The lexico-grammatical approach introduced by Mulder et al. (2004) focused on the propagation of affect towards an object. Boucouvalas (2003) developed the Text-to-Emotion Engine based on word tagging and analysis of sentences. The proposed system uses a small set of emotions, the six 'basic' types defined by Ekman (1993). The emotion extraction engine can analyse input text from a chat environment, identify the emotion communicated, and deliver the parameters necessary to invoke an appropriate expressive image on the user's display. However, the proposed system employs a parser that generates emotional output only if an emotional word refers to the person himself/herself and the sentence is in present continuous or present perfect continuous tense. We believe that such limitations greatly narrow the potential of textual emotion recognition. As a result, sentences like '*Onion pie is disgusting*' and '*It was the most joyous feeling!*' are disregarded by the parser despite the fact that they evidently carry affect. An approach for understanding the underlying semantics of language using large-scale real-world commonsense knowledge was proposed by Liu et al. (2003), who incorporated the affect sensing engine into an affectively responsive email composer called EmpathyBuddy.

There is a strong tie between our approach to affect analysis from text with the work of Moilanen and Pulman (2007) on sentiment composition. In their work, Moilanen and Pulman (2007) propose a theoretical composition model employing deep dependency parsing, sentiment propagation, polarity reversal, and polarity conflict resolution within various linguistic constituent types at various grammatical levels. The experiments with the developed lexical system revealed the crucial dependency on a wide-coverage lexicon, accurate parsing, and sentiment sense disambiguation in a compositional approach to sentiment analysis. The significant differences between our approaches lie in (1) the levels of classification (fine-grained classes, or nine emotions, and neutral, in our work versus polarity-based classes, or positive/negative/neutral in (Moilanen and Pulman 2007)); (2) the use of content words playing the role of reversers of a non-neutral polarity of related words (e.g., '*reduce*', '*prevention*') in their algorithm. However, it is not trivial to find pairs of opposite emotions in the case of a fine-grained classification, so such reversers cannot be applied here in a straightforward manner.

The text-based emotion prediction problem in the domain of children's fairy tales was explored by Alm et al. (2005) using supervised machine learning approach. As those researchers did not have sufficient training data to classify sentences according to fine-grained distinct emotions, in their preliminary study, Alm et al. (2005) focused only on three categories: neutral, positive emotion, and negative emotion. In her dissertation, Alm (2008) describes the refined and improved feature set, and presents the results of experiments on fine-grained emotion classification of text using hierarchical sequential model.

Style and level of formalism of written natural language differ greatly depending on situation. In news, reports, scientific papers etc., text is syntactically correct and written in a formal style, while in private correspondence, online messaging, and blogs, text is informal and may include special symbols, emoticons, abbreviations, and acronyms. The weakness of most affect recognition systems integrated with chat or e-mail browsers, however, is that they do not take into account crucial aspects of informal online conversation such as its specific style and evolving language. In order to account for the peculiarity of this medium, and to ensure satisfactory results on real examples, we investigated style, linguistic, and interactional features of online conversations (Neviarouskaya et al. 2007a), and took these into consideration in constructing our Affect Analysis Model.

3 Recognition of Fine-Grained Emotions from Text

In this section, we will summarize the main steps of emotion recognition using our Affect Analysis Model, which was introduced in (Neviarouskaya et al. 2007a, 2007b).

3.1 Basis for Affective Text Classification

For affective text classification, we decided to use the subset of emotional states defined by Izard (1977): 'anger', 'disgust', 'fear', 'guilt', 'interest', 'joy', 'sadness', 'shame', and 'surprise'. Besides emotions, we defined five communicative functions that are frequently observed in online conversations ('greeting', 'thanks', 'posing a question', 'congratulation', and 'farewell'). In this chapter we mainly focus on the recognition of emotional state, and do not discuss the task of detection of communicative functions.

In order to support the handling of abbreviated language and the interpretation of affective features of lexical items, the Affect database was created. The Affect database includes the following tables: Emoticons, Abbreviations, Adjectives, Adverbs, Nouns, Verbs, Interjections, and Modifiers. The affective lexicon was mainly taken from WordNet-Affect (Strapparava and Valitutti 2004). Emotion categories with intensities were manually assigned to the emotion-related entries of the database by three independent annotators. Emotion intensity values range from 0.0 to 1.0. Emoticons and abbreviations were transcribed and related to named affective states (with intensity), whereby each entry was assigned to only one category (e.g., emoticon ':-S' [worry] was related to 'fear' with intensity 0.4). Considering the fact that some affective words may express more than one emotional state, annotators could relate words to more than one category (e.g., the final annotation for noun '*enthusiasm*' is 'interest:0.8, joy:0.5'). Two annotators gave coefficients for intensity degree strengthening or weakening (from 0.0 to 2.0) to the adverbs of degree, and the result was averaged (e.g., coeff('*significantly*') = 2.0).

3.2 Affect Analysis Model

Our Affect Analysis Model was designed based on the compositionality principle (originally proposed as a method of semantic interpretation by Montague (1974)), according to which we determine the emotional meaning of a sentence by composing the pieces that correspond to lexical units or other linguistic constituent types governed by the rules of aggregation, propagation, domination, neutralization, and intensification, at various grammatical levels. By analysing each sentence in sequential stages (symbolic cue processing, detection and transformation of abbreviations, sentence parsing, and word/phrase/sentence-level analyses), our method is capable of processing sentences of different complexity, including simple, compound, complex (with complement and relative clauses), and complex-compound sentences.

3.2.1 Symbolic Cue Analysis

In the *first stage* of the Affect Analysis Model, we test the sentence for emoticons, abbreviations, interjections, '?' and '!' marks, repeated punctuation, and capital letters. Several rules are applied to define the dominant emotion in cases when multiple emoticons and emotion-relevant abbreviations occur in a sentence. As interjections are added to sentences to convey emotion (e.g. '*Oh no*', '*wow*'), they are analysed as well. If there are no emotion-relevant emoticons or abbreviations in a sentence, we prepare the sentence for parser processing: emoticons and abbreviations relating to communicative function categories are excluded from the sentence; and non-emotional abbreviations are replaced by their proper transcriptions found in the database (e.g., '*I m* [am] *stressed bc* [because] *i have frequent headaches*'). In such a way, the issue of correct processing of abbreviated text by syntactic parser is resolved.

3.2.2 Syntactic Structure Analysis

The *second stage* is devoted to the analysis of syntactic structure of sentences, and it is divided into two main subtasks. First, sentence analysis based on the Connexor Machinese Syntax[1] returns word base forms (lemmas), parts of speech, dependency functions representing relational information between words in

[1] Connexor Machinese Syntax. http://www.connexor.eu/technology/machinese/machinesesyntax/

sentences, syntactic function tags, and morphological tags. Second, parser output processing is performed. When handling the parser output, we represent the sentence as a set of primitive clauses (either independent or dependent). Each clause might include Subject formation (SF), Verb formation (VF) and Object formation (OF), each of which may consist of a main element (subject, verb, or object) and its attributives and complements. For the processing of complex or compound sentences, we build a so-called 'relation matrix', which contains information about dependences that the verbs belonging to different clauses have.

3.2.3 Word-Level Analysis

In the *third stage*, for each word found in our Affect database, the affective features of a word are represented as a vector of emotional state intensities e=[anger, disgust, fear, guilt, interest, joy, sadness, shame, surprise] (e.g., e('*remorsefully*')=[0,0,0,0.8,0,0,0.5,0,0]). In the case of a modifier, the system identifies its coefficient (e.g., coeff('*barely*')=0.4). Our model also varies the intensities of emotion vectors of adjectives and adverbs in comparative or superlative forms (e.g., e('*glad*')=[0,0,0,0,0,0.4,0,0,0], e('*gladder*')= [0,0,0,0,0,0.48,0,0,0] and e('*gladdest*')=[0,0,0,0,0,0.56,0,0,0]).

3.2.4 Phrase-Level Analysis

The purpose of this stage is to detect emotions involved in phrases, and then in Subject, Verb, or Object formations. We have defined rules for processing phrases:

1. Adjective phrase: modify the vector of adjective (e.g., e('*extremely doleful*') = coeff('*extremely*') * e('*doleful*') = 2.0 * [0,0,0,0,0,0,0.4,0,0] = [0,0,0,0,0,0,0.8,0,0]).
2. Noun phrase: output vector with the maximum intensity within each corresponding emotional state in analysing vectors (for instance, e1=[0..0.7..] and e2=[0.3..0.5..] yield e3=[0.3..0.7..]).
3. Verb plus adverbial phrase: output vector with the maximum intensity within each corresponding emotional state in analysing vectors (e.g., e('*shamefully deceive*')=[0,0.4,0,0,0,0,0.5,0.7,0] where e('*shamefully*')=[0,0,0,0,0,0,0,0.7,0] and e('*deceive*')= [0,0.4,0,0,0,0,0.5,0,0]).
4. Verb plus noun phrase: if verb and noun phrase have opposite polarities (e.g., '*break favourite vase*', '*enjoy bad weather*'), consider vector of verb as dominant; if polarities are the same (e.g., '*like honey*', '*hate crying*'), output vector with maximum intensity in corresponding emotional states.
5. Verb plus adjective phrase (e.g., '*is very kind*', '*feel bad*'): output vector of adjective phrase.

The rules for modifiers are as follows: (1) adverbs of degree increase or decrease emotional intensity values; (2) negation modifiers such as '*no*', '*never*', '*nothing*' etc. cancel (set to zero) vectors of the related words, i.e., 'neutralize the emotional content' (e.g., '*Yesterday I went to a party, but nothing exciting happened there*'); (3) prepositions such as '*without*', '*except*', '*against*', '*despite*' cancel vectors of related words (e.g., '*I climbed the mountain without fear*' is neutralized due to preposition).

Statements beginning with words like '*think*', '*believe*', '*sure*', '*know*', '*doubt*' or with modal verbs such as '*can*', '*may*', '*would*' etc. are neutralized by our system. Conditional clause phrases beginning with '*even though*', '*if*', '*unless*', '*whether*', '*when*', etc. are neutralized as well (e.g., '*I eat when I'm angry, sad, bored...*').

Each of the Subject, Verb, or Object formations may contain words conveying emotional meaning. During this stage, we apply the described rules to phrases detected within formation boundaries. Finally, each formation can be represented as a unified vector encoding its emotional content.

3.2.5 Sentence-Level Analysis

The emotion vector of a simple sentence (or a clause) is generated from Subject, Verb, and Object formation vectors resulting from phrase-level analysis. The main idea here is to first derive the emotion vector of Verb-Object formation relation. It is estimated based on the 'verb plus noun phrase' rule described above. In order to apply this rule, we automatically determine polarities of Verb and Object formations using their unified emotion vectors (particularly, non-zero-intensity emotion categories). The estimation of the emotion vector of a clause (Subject plus Verb-Object formations) is then performed in the following manner: (1) if polarities of Subject formation and Verb formation are opposite (e.g., SF = '*my darling*', VF = '*smashed*', OF '*his guitar*'; or SF = '*troubled period*', VF = '*luckily comes to an end*'), we consider the vector of the Verb-Object formation relation as dominant; (2) otherwise, we output the vector with maximum intensities in corresponding emotional states of vectors of Subject and Verb-Object formations.

To estimate the emotion vector of a compound sentence, first, we evaluate the emotion vectors of its independent clauses. Then, we define the resulting vector of the compound sentence based on two rules: (1) with comma and coordinate connectors 'and' and 'so' (e.g., '*It is my fault, and I am worrying about consequences*', '*The exotic birds in the park were amazing, so we took nice pictures*'), or with a semicolon with no conjunction: output the vector with the maximum intensity within each corresponding emotional state in the resulting vectors of both clauses; (2) with coordinate connector 'but' (e.g., '*They attacked, but we luckily got away!*'): the resulting vector of a clause following after the connector is dominant.

In order to process a complex sentence with a complement clause (e.g., '*I hope that Sam will not yell at my dog*'), first we derive the emotion vector of the complement clause, then create Object formation for the main clause using this vector, and finally estimate the resulting emotion vector of the main clause with added Object formation. In brief, we represent such sentence as a simple one, using the following pattern: 'who-subject does-verb what-object', where object is represented as a complement clause. In our algorithm, the complex sentences containing adjective (relative) clauses are analysed in the following manner: (1) the emotion vector of adjective clause is estimated; (2) this emotion vector is added to the SF or OF of the main clause depending on the role of the word to which the adjective clause relates; and (3) the emotion vector of the whole sentence is estimated.

While processing complex-compound sentences (e.g., '*Max broke the china cup, with which Mary was awarded for the best song, so he regretted profoundly*'), first we generate emotion vectors of dependent clauses, then of complex sentences, and finally, we analyse the compound sentence formed by the independent clauses. It is important to note that our system enables the differentiation of the strength of the resulting emotion depending on the tense of a sentence and availability of first person pronouns. The dominant emotion of the sentence is determined according to the emotion state with the highest intensity within the final emotion vector.

3.3 Evaluation of the Affect Analysis Model

3.3.1 Experiment with Blog Entries

To measure the accuracy of the proposed emotion recognition algorithm, we extracted 700 sentences from a collection of diary-like blog posts provided by BuzzMetrics[2]. Particularly, we focused on online diary or personal blog entries, which are typically written in a free style and are rich in emotional colourations. The most noticeable aspects of diary-like text are privacy, naturalism and honesty in the expression of the author's thoughts and feelings.

Three independent annotators labeled the sentences with one of nine emotion categories, or neutral, and a corresponding intensity value. For the evaluation of algorithm performance, we created two collections of sentences corresponding to different 'gold standards': (1) 656 sentences, on which two or three human raters completely agreed (Fleiss' Kappa coefficient is 0.51); (2) 249 sentences, on which all three human raters completely agreed (Fleiss' Kappa coefficient is 1.0). Table 1 shows the distributions of emotion labels across 'gold standard' sentences.

Table 1. The distributions of emotion labels across 'gold standard' sentences

Number of sentences with distinct emotion labels										
neutral	anger	disgust	fear	guilt	interest	joy	sadness	shame	surprise	total
'Gold standard' sentences, on which at least 2 annotators agreed										
75	59	30	49	22	43	181	145	9	43	656
'Gold standard' sentences, on which all 3 annotators agreed										
8	17	9	24	12	8	88	58	3	22	249

We developed two versions of the Affect Analysis Model (AAM) differing in syntactic parsers employed during the second stage of affect recognition algorithm: (1) AAM with commercial parser Connexor Machinese Syntax

[2] Weblog Data Collection. BuzzMetrics, Inc. http://www.nielsenbuzzmetrics.com

(AAM-CMS); (2) AAM with GNU GPL licensed Stanford Parser[3] (Klein and Manning 2003) (AAM-SP). The performance of the AAM-CMS and AAM-SP was evaluated against both sets of sentences related to 'gold standards'. Averaged accuracy, precision, recall, and F-score are shown in Table 2 for each fine-grained emotion category. Additionally, we provide the results for merged labels (positive emotions including 'interest', 'joy', and 'surprise'; negative emotions including 'anger', 'disgust', 'fear', 'guilt', 'sadness', and 'shame'; and neutral).

Table 2. The results of experiment with Affect Analysis Model employing different parsers

AAM version	Measure	Fine-grained emotion categories										Merged labels		
		neut	**ang**	**disg**	**fear**	**guilt**	**inter**	**joy**	**sad**	**sh**	**sur**	**Pos**	**Neg**	**Neut**
'Gold standard' sentences, on which at least 2 annotators agreed														
AAM-CMS	Accuracy	**0.726**										**0.816**		
	Precision	0.46	0.83	0.63	0.76	0.75	0.56	0.87	0.78	0.57	0.85	0.85	0.92	0.41
	Recall	0.55	0.41	0.73	0.84	0.68	0.88	0.83	0.72	0.89	0.77	0.90	0.81	0.55
	F-score	0.50	0.55	0.68	0.80	0.71	0.68	0.85	0.75	0.70	0.80	0.87	0.86	0.47
AAM-SP	Accuracy	**0.649**										**0.747**		
	Precision	0.30	0.77	0.64	0.74	0.71	0.61	0.83	0.74	0.50	0.76	0.84	0.91	0.28
	Recall	0.55	0.34	0.70	0.80	0.55	0.81	0.71	0.64	0.67	0.72	0.80	0.75	0.55
	F-score	0.39	0.47	0.67	0.76	0.62	0.70	0.76	0.69	0.57	0.74	0.82	0.82	0.37
'Gold standard' sentences, on which all 3 annotators agreed														
AAM-CMS	Accuracy	**0.815**										**0.890**		
	Precision	0.26	0.92	0.83	0.91	0.83	0.44	0.95	0.88	0.67	0.86	0.93	0.99	0.15
	Recall	0.75	0.65	0.56	0.88	0.83	0.88	0.88	0.79	0.67	0.82	0.94	0.86	0.75
	F-score	0.39	0.76	0.67	0.89	0.83	0.58	0.91	0.84	0.67	0.84	0.94	0.92	0.25
AAM-SP	Accuracy	**0.751**										**0.814**		
	Precision	0.15	0.92	0.83	0.87	0.80	0.50	0.96	0.88	0.50	0.82	0.94	0.98	0.08
	Recall	0.75	0.65	0.56	0.83	0.67	0.75	0.78	0.74	0.33	0.82	0.85	0.79	0.75
	F-score	0.24	0.76	0.67	0.85	0.73	0.60	0.86	0.80	0.40	0.82	0.89	0.88	0.14

The analysis of the failures of the Affect Analysis Model revealed that common sense or additional context is required for processing some sentences. For example, human annotators agreed on the 'sadness' emotion conveyed through '*What I hope is that he can understand how much I treasure this friendship*', while our lexicon-based system resulted in erroneous 'joy' emotion. This is due to the fact that this sentence contains two words annotated by 'joy' label in our Affect database, namely, '*hope*' and '*treasure*', but there are no lexical terms indicating 'sadness' emotion. In some cases, where system result did not agree with the 'gold standard' due to the rule of neutralization of negated phrases (e.g., sentence '*I*

[3] Stanford Parser. http://nlp.stanford.edu/software/lex-parser.shtml

don't care whether they like me at the cocktail parties, or not' was annotated by humans as expressing 'anger' emotion, and by our system as 'neutral'), the solution would be to reverse the polarity of a statement (e.g., positive '*care*' with negation should become negative phrase '*don't care*'), however, finding the pairs of opposite emotions might be problematic. Neutralizations due to 'cognition-related' ('*assume*', '*know*', '*think*'), modal ('*can*', '*could*', '*may*', '*would*'), and condition ('*if*') words also caused the problematic interpretations (e.g., AAM resulted in 'neutral' emotion in sentences '*I tried explaining to him my outlooks on life last night, and I think that I upset him*', '*She knows that she can trust me, I'll never do her wrong*', and '*And if he would laugh when it happens that would only make me more angry and thus blow up at him*', while 'gold standard' annotations were 'sadness', 'joy', and 'anger', correspondingly). Such results generate a need for more careful analysis of the cases where condition or modal operators are involved in the sentence. Other errors were caused by the lack of relevant terms in the Affect database (e.g., emotion in a sentence '*He's just lying*' was not recognized by our system as word '*lie*' was not included in the lexicon), incorrect results from syntactic parser, and sense ambiguity.

It is worth noting, however, that the accuracy of the Affect Analysis Model with the (commercially available) parser (Connexor Machinese Syntax) was higher in 6-8 % on the same sets of sentences (Table 2). This indicates that Stanford Parser employed for the syntactic structure analysis is less efficient. On the other hand, as we aim to freely distribute and apply our emotion recognition tool to textual messages in online environments, we have to compromise on the performance of the system for the sake of free distribution.

3.3.2 Experiment with Sentences from Fairy Tales

In our second experiment, we decided to compare the performance of the AAM (employing Connexor Machinese Syntax) with Alm's (2008) system that outperformed Liu's (2003) system on sentences from fairy tales. Following the same evaluation scenario as Alm (2008), we considered three hierarchical levels of affect labels: TOP (emotional, neutral); MID (positive, negative, neutral); and ALL (merged 'joy'-'interest', 'surprise' under positive; merged 'anger'-'disgust', 'fear', merged 'sadness'-'guilt'-'shame' under negative; and neutral).

We ran the experiment on a subset of 1207 sentences marked by high agreement[4] (affect labels assigned by four human annotators were identical), and a subset of sentences with neutral label. As we did not have the subsets of neutral sentences used by Alm in her experiments, we randomly extracted them from the whole corpus of sentences that were labelled by human annotators as neutral. The size of the sample of neutral sentences varied at each hierarchical level and was determined based on the number of affective labels at each level by Equation 1 (taken from (Alm 2008)).

$$\left\lfloor \frac{|HA|}{|Ai|-1} \right\rfloor, \qquad (1)$$

[4] Affect data from fairy tales. http://lrc.cornell.edu/swedish/dataset/affectdata/index.html

where *HA* is the set of high agreement affect sentences; *Ai* is the set of affect labels at a level *i* in the affect hierarchy.

We compared the results of the AAM with Alm's (2008) LOOHAsnowtag method (supervised machine learning approach) and two baselines (Table 3). The ratio of neutrally labelled sentences was considered as first baseline (N-BL), while the second baseline (Freq-BL) was represented by the ratio of most frequent affect label.

Table 3. Accuracy across sentences from fairy tales in high agreement experiment (span of mean accuracy given for LOOHAsnowtag method)

	Data size	**Baselines**		**Individual classification methods**			
Level	**Total # of sentences (# of N sentences)**	**N-BL**	**Freq-BL**	**LOOHA-snowtag**	**AAM**	**AAM_LEX**	**AAM_LEX_WI**
ALL	1448 (241)	17	31 (J-I)	69-70	68.2	70.1	**70.2**
MID	1810 (603)	33	40 (NEG)	69-73	73.3	74.6	**75.7**
TOP	2414 (1207)	50	50 (any)	79	77.6	78.5	**79.9**

As seen from Table 3, AAM resulted in similar accuracy level as LOOHAsnowtag method, outperforming both baselines. Since our approach is completely lexical, and its recall depends on the coverage of the lexicon used (about 7 % of the errors in high agreement affect sentences at the ALL level were due to the lack of affect entries in database), we improved the performance of our system (see results of AAM_LEX in Table 3) by expanding the Affect database with some sample words from experimental data. The decision for further improvement (see results of AAM_LEX_WI in Table 3, where WI means 'without Interest' or 'neutral label instead of Interest') was based on the fact that about 8.5 % of errors in neutral sentences at the TOP level were comprised of the cases where AAM system resulted in 'Interest' emotion that was not considered as a label in the Alm's (2008) set of affect labels.

In summary, the evaluation of the Affect Analysis Model algorithm showed promising results regarding its capability to recognize affective information in sentences from an existing corpus of informal online communication and fairy tales. However, the current main limitations of the developed affect recognition module must be noted, namely: strong dependency on the source of lexicon, Affect database, and the syntactic parser; no disambiguation of word senses; and disregard of contextual information.

4 AffectIM: An Emotionally Intelligent Instant Messaging System

To enrich the user's experience in online communication, to make it enjoyable, exciting and fun, we implemented a web-based Instant Messaging (IM) application,

AffectIM (Neviarouskaya et al. 2008), and endowed it with emotional intelligence by integrating the developed Affect Analysis Model.

4.1 System Architecture and User Interface

In AffectIM, we propose to equip the user with an avatar, i.e. a graphical representation of the user in Instant Messaging. The avatar is endowed with the ability to express emotions and to exhibit social nonverbal behavior, whereby its behavior is based on textual affect sensing, and the interpretation of communicative functions conveyed by online conversations. The system architecture is schematically represented in Fig. 1.

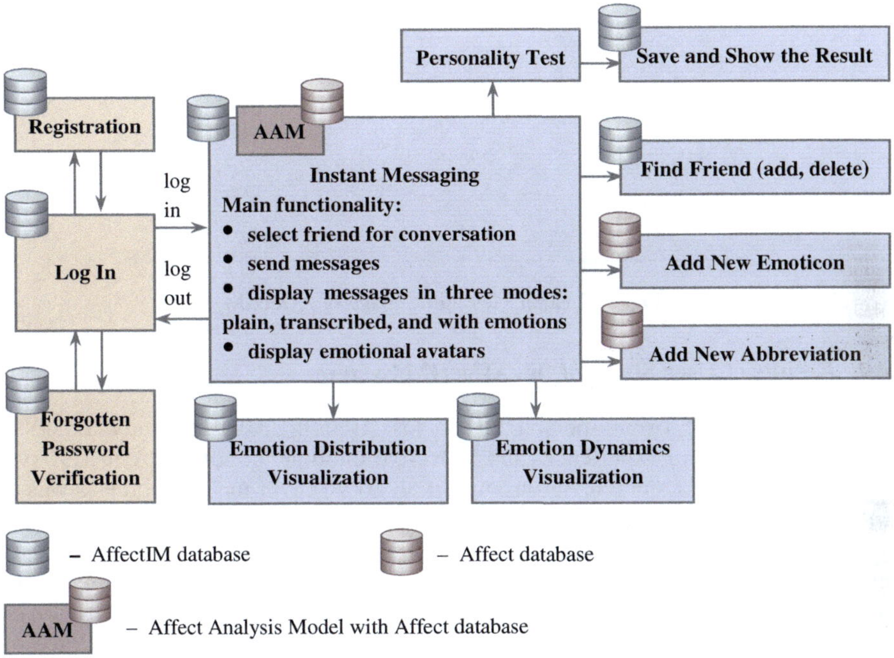

Fig. 1. Architecture of the AffectIM system

Within our time-limited research project, we could design only two avatars, one male and one female. So the graphical representative is automatically selected by the system according to the user's sex. The main window of the AffectIM system, showing an online conversation, is depicted in Fig. 2.

From the list of friends displayed in the left frame, the user selects the person (available online), whom he or she wishes to communicate with. The central frame allows the user to type and to send the messages. It displays the conversation flow in three modes: plain, transcribed, and with emotions. Further, it displays emotional avatars (own – to the left of conversation field, and friend's – to the right). Two buttons located under the avatar animation refer to the visualization of emotion distribution (either in a color bar or pie graph) and emotion dynamics (line graph). Since the language of online communication is constantly evolving, AffectIM also provides the functionality to add new abbreviations, acronyms, and emoticons to the Affect database (see two buttons located to the left from the input text field in Fig. 2).

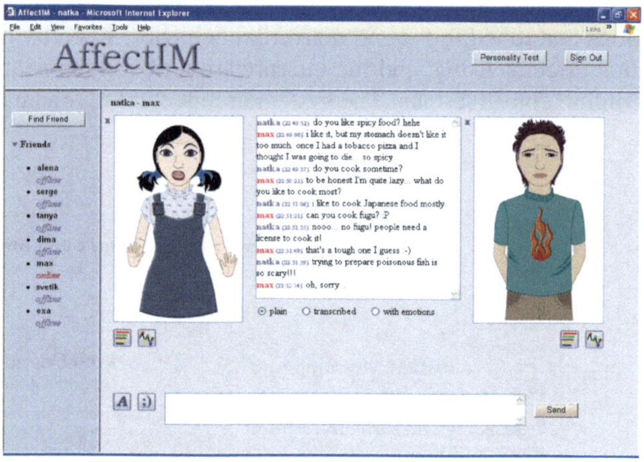

Fig. 2. 'Instant Messaging' page of AffectIM

4.2 Results of User Study of the AffectIM System

We conducted the twenty-person (10 males, 10 females) user study aimed to evaluate 'richness of experience' and 'affective intelligence' of our AffectIM system (details are given in (Neviarouskaya et al. 2008)). Our main hypothesis was as follows: '*The user experience and effectiveness of communication will be better when the interface features emotionally expressive avatars, which convey user emotions that are automatically extracted from their instant messages, than when using an interface where the user has to select avatar emotions manually, or an interface with randomly displayed avatar emotions.*' Our hypothesis was tested by considering the following dimensions regarding users' experience:

1. Interactivity.
2. Involvement (engagement).
3. Sense of copresence.
4. Enjoyment.
5. Affective intelligence.
6. Overall satisfaction.

In addition to these main criteria, we asked participants to give us feedback on some general questions.

The experiment was designed as a within-subjects experiment in pairs. In particular, we compared three AffectIM interfaces using different configuration conditions. For the user study, we prepared three versions of the system:

1. Automatic (A-condition). In this interface, affect sensing from text is performed based on the developed Affect Analysis Model, and the recognized emotions are conveyed by the avatar expressions.
2. Manual (M-condition). During this condition, no automatic emotion recognition from text is performed; however, users may select emotion (and its intensity) to be shown by avatars using 'select pop-up menus'.
3. Random (R-condition). Here, the avatars show a 'quasi-random' reaction. First, we process each sentence using the Affect Analysis Model, and then we apply two rules: (1) if the output is emotional, we run two functions that randomly select the emotion out of nine available emotions and its intensity, correspondingly; (2) for the case of 'neutral' output, we set the function that generates 'neutral' emotion with the probability of 60 % or 'random' emotion with the probability of 40 %.

The average duration of sessions on each interface was 10.1 minutes (minimum 8 and maximum 12.5 minutes), excluding the time needed to fill out the questionnaires. The 11 questions on main criteria were answered based on 7-item agreement Likert scale. Since our study involved each subject being measured under each of three conditions (within-subjects design), we analyzed the data using statistical method ANOVA (an extension of matched pair t-test). The list of questions according to the analyzed criteria, mean scores, and ANOVA results (p-values) are summarized in Table 4.

The results on *interactivity* showed that subjects tended to consider the condition, which allowed them to manipulate the expressed emotion manually, as most interactive. However, ANOVA resulted in no significant difference in interactivity among three interfaces. The reported *involvement* of all three systems also did not differ significantly, showing that the level of engagement was almost the same. The following two questionnaire items covering the aspects of space and togetherness are intended for evaluation of *sense of copresence*, or social presence: '*I felt if I were communicating with my conversation partner in the shared virtual space*', '*The system gave me the sense that the physical gap between us was narrowed*'. The statistic ANOVA results for the first questionnaire item indicated the significance of the difference in sense of copresence felt in A-condition and R-condition (p (R-A) < 0.05), and showed that the A-condition gave stronger feeling of communication in the shared virtual space than the R-condition. No significant difference among three interfaces was reported on the second statement. The high levels of *enjoyment* were reported during A-condition and M-condition.

To evaluate *affective intelligence*, four statements (three – directly related to the system and one – indirectly related) were proposed to subjects in questionnaire: '*The system was successful at conveying my feelings*' (Q1), '*The system was successful at conveying my partner's feelings*' (Q2), '*The emotional behavior of the avatars was appropriate*' (Q3), and '*I understood the emotions of my communication partner*' (Q4).

The systems in A-condition and M-condition (with small prevalence of mean results in M-condition) were both more successful at conveying own feelings (see Q1) than the system in R-condition. Since M-condition is considered as a 'gold standard' in communicating person's emotions, and ANOVA showed no

Table 4. The list of questions, mean scores, and ANOVA results (p-values)

Criteria	Questionnaire items	Average scores			p-values for pairs of conditions		
		R*	**M***	**A***	**R-M**	**R-A**	**M-A**
Interactivity	*The system was interactive*	4.7	5.5	5.2	0.068	0.106	0.356
Involvement (engagement)	*I felt it was important for my conversation partner that I responded after each his/her statement*	5.5	5.35	5.15	0.691	0.376	0.551
	I was awaiting the replies of my conversation partner with true interest	5.45	5.45	5.45	1.0	1.0	1.0
Sense of copresence	*I felt if I were communicating with my conversation partner in the shared virtual space*	4.25	4.6	4.95	0.273	**0.023**	0.232
	The system gave me the sense that the physical gap between us was narrowed	3.8	4.3	4.15	0.154	0.260	0.659
Enjoyment	*I enjoyed the communication using this IM system*	4.5	4.85	4.8	0.246	0.301	0.881
Affective intelligence	*The system was successful at conveying my feelings*	3.4	4.5	4.45	**0.010**	**0.008**	0.883
	The system was successful at conveying my partner's feelings	4.05	4.6	4.35	0.086	0.368	0.262
	The emotional behavior of the avatars was appropriate	3.3	4.5	4.6	**0.004**	**0.004**	0.772
	I understood the emotions of my communication partner	4.25	4.6	4.35	0.297	0.789	0.349
Overall satisfaction	*I am satisfied with the experience of communicating via this system*	4.25	4.6	4.6	0.232	0.309	1.0

*R, M, and A stand for Random, Manual, and Automatic conditions, correspondingly.

significant difference between M-condition and A-condition, we might say that automatic emotion recognition system performed well enough to bring high affective intelligence to IM application. As was expected, significant differences were found between R-condition and M-condition (p (R-M) < 0.05), and between R-condition and A-condition (p (R-A) < 0.01).

While evaluating successfulness of the interfaces at conveying conversation partner's feelings (see Q2), the highest rate was given by subjects to M-condition, and the lowest – to R-condition. However, ANOVA for this criterion resulted in no significant difference among all interfaces. One user's comment regarding the emotional reactions of the partner's avatar was: 'I concentrated too much on the reactions of my avatar and not enough on that of my partner. Reading and thinking about the answer took away the concentration on the avatar'.

Interesting results were observed for the evaluation of appropriateness of emotional behavior of avatars (see Q3). Results for A-condition and M-condition significantly prevailed those for R-condition ($p\ (R\text{-}A) < 0.01$; and $p\ (R\text{-}M) < 0.01$). Users' comments confirmed that during R-condition subjects sometimes couldn't understand why the avatars did not correspond to their words and reacted in 'wrong' ways. Although A-condition was rated a little bit higher than M-condition, no significant difference was detected between these interfaces.

The Q4 statement measured affective intelligence of the system indirectly, since people used to derive emotional content from text based on semantic information and their empathetic abilities. Emotional expressions of avatars may help to understand the partner's emotion clearer. As was expected, the highest rate was reported in M-condition, and the lowest – in R-condition, where participants might be confused, since sometimes emotions shown by the avatar contradict actual emotional content. However, no significant difference was found in partner's emotion comprehension among all three interfaces. A possible explanation for such results might be that a person typically relies on his/her own affective intelligence rather than on results of artificial affective intelligence.

The *overall satisfaction* from using three AffectIM interfaces was evaluated using statement '*I am satisfied with the experience of communicating via this system*'. Regarding the results, average scores for A-condition and M-condition were equal (4.6), whereas less satisfaction was reported for R-condition (4.25).

After measuring the overall mean results of scores given by males and females separately, we found out that, in general, females gave higher scores to the questionnaire statements than males (average score difference was 0.65). This might indicate that female subjects were more positive in their judgments.

In addition to the main questionnaire items, after finishing communications through all three interfaces, participants were given general questions. Subjects were asked to associate nine emotion states with nine avatar expressions shown on still figures. The detected pairs of most often confused emotions are 'fear' – 'surprise' and 'guilt' – 'sadness'; and less often confused emotions are 'guilt' – 'fear' and 'sadness' – 'fear'. Some participants confused emotions in 'interest' – 'joy' and 'surprise' – 'guilt' pairs. These results suggest that during the experiment some participants might have faced a difficulty with correct interpretations of emotional behavior of avatars. (We say 'might' because emotional avatar expression in IM always occurs in the context of the textual message, which contributes to the disambiguation of the expression.)

The participants' comments and the results of answers to the question '*To what degree do you think is necessary to look at a graphical representation of the other communicating person?*' suggest that there are two types of IM users: (1) some are open to new features of IM, and find animated graphical representation of a person helpful in understanding the partner's emotions and giving some sense of physical presence; (2) others tend to concentrate their attention on content, and prefer small emotional symbolic cues, like emoticons, to avatar expressions.

The user opinions regarding the 'select emotion' menu feature were diverse. Some users criticized the type of pop-up menu, commenting that it was difficult to use, it took long time to select, and choosing emotion intensity on fine-grained scale was overwhelming. For more convenience, they proposed to replace pop-up menus by icons and spread them out. One of the subjects complained that emotion select menu disturbed the flow of the chat. Another reported problem is that since there is no preview of what the emotion expression looks like, it is unclear whether it matches the user's intention. Some subjects felt that basic emotions are too general and are not sufficient to convey emotion in many cases (like, for example, in phrase '*hey, just kidding*', where users typically use emoticon). Also, they suggested providing the possibility of showing more different or even mixed emotions (some state between sadness and joy). However, we think that displaying mixed emotional expressions would add more confusion and misinterpretation to the conversations.

Some subjects underlined positive aspects of manual selection of emotion states. They found this function interesting and helpful, because (1) it offered the possibility to visually express feelings and better understand them, (2) it allowed preventing inappropriate emotional reaction of the avatar, and (3) guaranteed accuracy of communicated emotion. This indicates that in sensitive conversations users would prefer manual control to avoid system mistakes that could sometimes harm the conversation.

Regarding the users' general impressions of using the AffectIM system, they found automatic recognition and visual representation of emotions to be a good, promising idea and fun to watch. The representation of emotions through the avatars was considered interesting, clear and easy to understand. However, some participants reported the difficulty of distinguishing the displayed emotions and stated that the system would benefit from more customizable avatars or even abstract avatars like smileys. Several users commented that they would also like to express a wider range of emotions.

From the experiment we learned that the IM application might benefit from an integration of automatic emotion sensing with manual control of emotional behavior of avatars in one interface, which will allow users to select between two modes depending on type and sensitivity of conversation.

5 EmoHeart: Automation of Expressive Communication of Emotions in Second Life

The 3D virtual world of 'Second Life' imitates a form of real life by providing a space for rich interactions and social events. Second Life encourages people to establish or strengthen interpersonal relations, to share ideas, to gain new experiences, and to feel genuine emotions accompanying all adventures of virtual reality. Emotional expression is natural and very important for communication in real life, but currently rather cumbersome in Second Life, where expressions have to be selected and activated manually. Concretely, a user has to click on the animation gesture in the list, or type the predefined command following the symbol '/' in a textual chat entry. In order to breathe emotional life into graphical representations of users (avatars) through the automation of emotional expressiveness, we applied the developed Affect Analysis Model to textual chat in Second Life (Neviarouskaya et al. 2010). The architecture of the system is presented in Fig. 3.

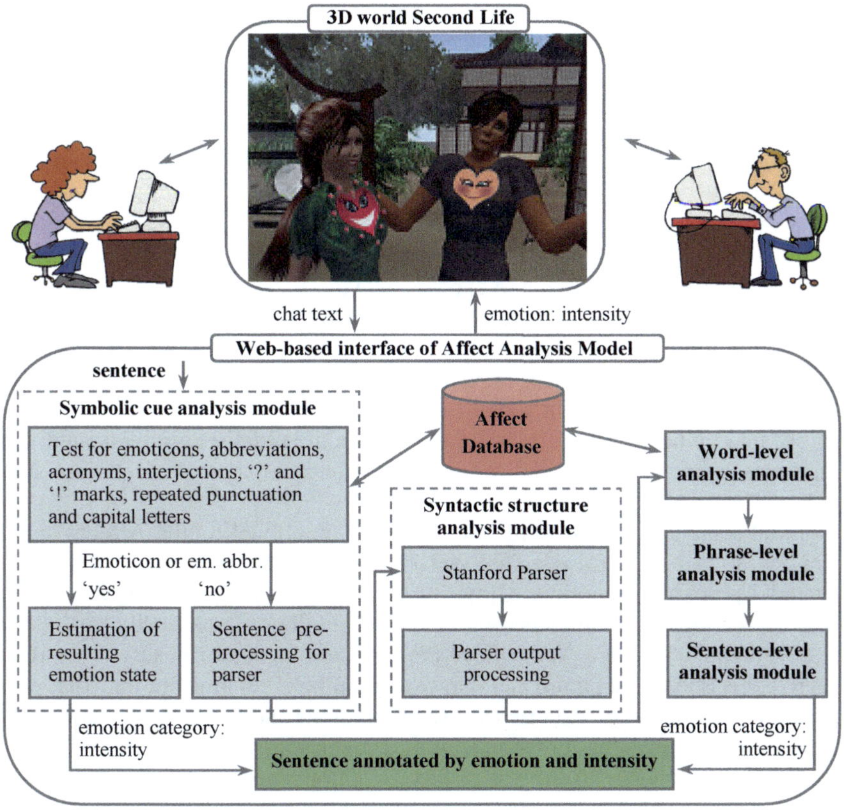

Fig. 3. Architecture of the EmoHeart system

The control of the conversation is implemented through the Second Life object called EmoHeart[5] (invisible in case of 'neutral' state) attached to the avatar's chest. The distributor of the EmoHeart object is located inside a (fictitious) Starbucks cafe[6] of the Second Life replica of National Center of Sciences building in Tokyo, which also hosts the National Institute of Informatics (NII). Once attached to the avatar, the EmoHeart object (1) listens to each message of its owner (i.e., user of the Second Life avatar), (2) sends it to the web-based interface of the Affect Analysis Model located on the server, (3) receives the result (dominant emotion and intensity), and (4) visually reflects the sensed affective state through (i) the animation of avatar's facial expression, (ii) EmoHeart texture (indicating the type of emotion), and (iii) size of the texture (indicating the strength of emotion,

[5] EmoHeart. http://www.prendingerlab.net/globallab/?page_id=22
[6] Second Life landmark. http://slurl.com/secondlife/NIIsland/213/38/25/

namely, 'low', 'middle', or 'high'). If no emotion is detected in the text, the EmoHeart remains invisible and the facial expression remains neutral.

Of the bodily organs, the heart plays a particularly important role in our emotional experience. People often characterize personal traits, emotional experiences, or mental states using expressions originating from word '*heart*' (for ex., '*heart-felt*', '*warm-hearted*', '*heartlessness*', '*kind-heartedness*', '*broken-hearted*', '*heart-burning*', '*heart-to-heart*' etc.). The essence of emotional, moral, and spiritual aspects of a human being has long been depicted using heart-shaped symbol. With the heart-shaped object of EmoHeart, we provide an additional channel for visualizing emotions in a vivid and expressive way.

We made EmoHeart available for Second Life users from December 2008. During a two month period (December 2008 – January 2009), we asked students to promote the EmoHeart object by visiting locations in Second Life and engaging other Second Life residents in social communication. As a result, 89 Second Life users became owners of EmoHeart, and 74 of them actually communicated using it. Text messages along with the results from Affect Analysis Model were stored in an EmoHeart log database.

From all sentences, 20 % were categorized as emotional by the Affect Analysis Model and 80 % as neutral (Fig. 4). We observed that the percentage of sentences annotated by positive emotions ('joy', 'interest', 'surprise') essentially prevailed (84.6 %) over sentences annotated by negative emotions ('anger', 'disgust', 'fear', 'guilt', 'sadness', 'shame'). We believe that this dominance of positivity expressed through text is due to the nature and purpose of online communication media, which allows people to exchange experiences, share opinions and feelings, and satisfy their social need of interpersonal communication. Harker and Keltner (2001) empirically verified that the tendency to express positive emotions creates more harmonious social relationships, which in turn fosters personal growth and well-being.

We analysed the distribution of emotional sentences from EmoHeart log data according to the fine-grained emotion labels from our Affect Analysis Model. We found that the most frequent emotion conveyed through text messages is 'joy' (68.8 % of all emotional sentences), followed by 'surprise', 'sadness' and 'interest' (9.0 %, 8.8 %, and 6.9 %, respectively). All remaining emotions individually do not exceed the level of 2.1 %. The least frequent emotion detected from text messages is 'shame' (0.6 % of all emotional sentences).

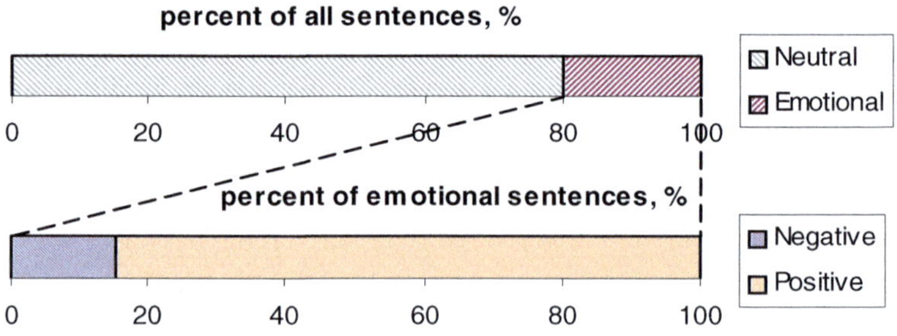

Fig. 4. Percentage distribution of emotional (positive or negative) and neutral sentences

As the Affect Analysis Model also enables detection of five communicative functions (besides nine distinct affective states) that are frequently observed in online conversations, we analysed the communicative functions identified in the EmoHeart log data as well. Our observations suggest that during online chatting people often ask each other questions (60.5 % of the cases of detected communicative functions), requesting thus for new information, confirmation or denial. Such social behaviours as 'greeting' and 'farewell', that are constituent parts of face-to-face communication, were recognized in 21.3 % and 7.3 % of the cases, respectively. EmoHeart users expressed gratitude in 10.7 % and congratulate each other in 0.2 % of the cases of detected communicative functions.

6 iFeel_IM!: System for Affective Haptic Communication

Driven by the motivation to enhance emotionally immersive experience of real-time messaging, we developed the system iFeel_IM! (**i**ntelligent system for **Feel**ing enhancement powered by affect sensitive **I**nstant **M**essenger), the philosophy behind of which is "***I feel*** [therefore] ***I am!***" (details are given in (Tsetserukou et al. 2009)). This system integrates (1) Second Life (with EmoHeart) as a medium for communication, (2) Affect Analysis Model, and (3) innovative affective haptic interfaces providing additional nonverbal communication channels through simulation of emotional feedback and social touch (physical co-presence).

We selected four distinct emotions that have strong physical features: ('anger', 'fear', 'sadness', and 'joy') for presentation through haptic devices. The precision of AAM in recognition of these emotions (see Table 2) is considerably higher ('anger' – 92 %, 'fear' – 91 %, 'joy' – 95 %, 'sadness' – 88 %) than of other emotions. The developed affective haptic devices make up three groups. The first group is intended for emotion elicitation implicitly (HaptiHeart, HaptiButterfly, HaptiTemper, and HaptiShiver), the second type evokes affect in a direct way (HaptiTickler), and the third one uses sense of social touch (HaptiHug) for influencing the mood and providing some sense of physical co-presence. Based on the emotion or 'hugging' cue detected in text, the corresponding haptic device worn by a user during remote communication is activated.

7 Conclusions

This chapter introduced a rule-based syntactic approach to affect recognition from text messaging. The purpose of our work is to improve expressiveness and (emotional) intelligence of computer-mediated communication. For textual input processing, our Affect Analysis Model handles not only correctly written text, but also informal messages written in an abbreviated or expressive manner. The salient features of the Affect Analysis Model are the following:

1. Analysis of nine emotions on the level of individual sentences: this is an extensive set of labels if compared to six emotions used in most of the related work.

2. The ability to handle the evolving language of online communications: to the best of our knowledge, our approach is the first attempt to deal with informal and abbreviated style of writing, often accompanied by the use of emoticons.
3. Foundation in database of affective words (each term in our Affect database was assigned at least one emotion label along with emotion intensity, in contrast to annotations of one emotion label or polarity orientation in competing approaches), interjections, emoticons, abbreviations and acronyms, modifiers (which influence on degrees of emotion states).
4. Vector representation of affective features of words, phrases, clauses, and sentences.
5. Consideration of syntactic relations and semantic dependences between words in a sentence: our rule-based method accurately classifies context-dependent affect expressed in sentences containing emotion-conveying terms, which may play different syntactic and semantic roles.
6. Analysis of negation, modality, and conditionality: most researchers ignore modal expressions and condition prepositions; therefore, their systems show poor performance in classifying neutral sentences, which is indeed not an easy task.
7. Consideration of relations between clauses in compound, complex, or complex-compound sentences: to our knowledge, AAM is the first system to comprehensively process affect reflected in sentences of different complexity.
8. Emotion intensity estimation: in our work, the strength of emotion is encoded through numerical value in the interval [0.0, 1.0], in contrast to low/middle/high levels detected in some of competing methods.

Our system showed promising results in affect recognition of real examples of online conversation. On sentences from 'gold standard' where at least 2 annotators agreed, accuracy was 73 % for fine-grained emotion classification, and 82 % for polarity-based merged categories; on sentences from 'gold standard' where all 3 annotators agreed, accuracy was 82 % for fine-grained emotion classification, and 89 % for polarity-based merged categories. Furthermore, for sentences extracted from fairy tales the averaged accuracy was 70 % for fine-grained emotion classification, and 76 % for polarity-based merged categories.

Currently, the main limitations of the developed affect recognition module are: strong dependency on the source of lexicon, Affect database, and the commercially available syntactic parser; no disambiguation of word meanings; disregard of contextual information and conversation history; and inability to recognize and process misspelled words in a sentence. In our future study we will investigate those issues and explore the possibilities to overcome the current limitations of the system. For example, Affect database can be expanded automatically by considering (1) direct synonymy (e.g., '*congratulate*' – '*felicitate*') and antonymy (e.g., '*brave*' – '*faint-hearted*') relations, and hypernym-hyponym (e.g., '*success*' – '*winning*') relations in WordNet (Miller 1990); (2) compounding using known emotion-carrying base components (e.g., '*well-wishing*', '*ill-conditioned*', '*terror-haunted*'); (3) morphologic modifications (e.g., '*harmony*'+'-*ous*'=>'*harmonious*', '*dis-*'+ '*honest*'=>'*dishonest*'). As our system is completely lexical and the language of online conversations is 'evolving', in the future we are planning to realize a procedure for the automatic updating of the Affect database. In addition, we will thoroughly examine different cases of the use of modals and condition prepositions in sentences, so that to be able to automatically distinguish emotional cases from neutral ones (in contrast to simple neutralization applied in current work). With respect to the rules for composition of emotion vectors of terms comprising phrases or clauses, we believe the approach aiming at learning rules from corpora would be useful.

We implemented a web-based IM application, AffectIM, and endowed it with emotional intelligence by integrating the Affect Analysis Model. The user study conducted on AffectIM showed that the IM system with automatic emotion recognition function was successful at conveying users' emotional states during communication online, thus enriching expressivity and social interactivity of online communications. From the experiment we learned that the IM application might benefit from an integration of automatic emotion sensing with manual control of emotional behavior of avatars in one interface, which will allow users to select between two modes depending on type and sensitivity of conversation.

This chapter also summarized the integration of the developed emotion recognition module into the 3D virtual world Second Life, where the Affect Analysis Model serves as the engine behind automatic visualization of emotions conveyed through textual messages. The control of the conversation in Second Life is implemented through the EmoHeart object attached to the avatar's chest. This object communicates with Affect Analysis Model located on the server, and visually reflects the sensed affective state through the animation of avatar's facial expression, EmoHeart texture, and size of the texture. In the future, we aim to study cultural differences in perceiving and expressing emotions, and to integrate a text-to-speech engine with emotional intonations into textual chat of Second Life.

While developing the iFeel_IM! system, which was shortly summarized in the chapter, we attempted to bridge the gap between mediated and face-to-face communications by enabling and enriching the spectrum of senses such as vision and touch along with cognition and inner personal state. Our primary goal for the future research is to conduct extensive user study on iFeel_IM! system.

Acknowledgments. We would like to acknowledge and thank Alessandro Valitutti and Dr. Diego Reforgiato for their kind help during the Affect database creation. We wish also to express our gratitude to Dr. Dzmitry Tsetserukou, Dr. Shaikh Mostafa Al Masum, Manuel M. Martinez, Zoya Verzhbitskaya, Hutchatai Chanlekha, and Nararat Ruangchaijatupon who have contributed to annotations of Affect database entries and sentences, for their efforts and time. We would like to thank Dr. Ulrich Apel and Dr. Boris Brandherm for their efforts and help in organization of the user study on AffectIM. Special thanks also go to Cui Xiaoke, Tananun Orawiwattanakul, and Farzana Yasmeen for their work on EmoHeart promotion in Second Life. This research was partly supported by a JSPS Encouragement of Young Scientists Grant (FY2005-FY2007), an NII Joint Research Grant with the University of Tokyo (FY2007), and an NII Grand Challenge Grant (FY 2008).

References

Allwood, J.: Bodily communication dimensions of expression and content. In: Multimodality in Language and Speech Systems, pp. 7–26. Kluwer Academic Publishers, Netherlands (2002)

Alm, C.O.: Affect in text and speech. Ph.D. Dissertation, University of Illinois at UrbanaChampaign (2008)

Alm, C.O., Roth, D., Sproat, R.: Emotions from text: machine learning for text-based emotion prediction. In: Proceedings of Human Language Technology Conference and Conference on Empirical Methods in Natural Language Processing, pp. 579–586 (2005)

Andreevskaia, A., Bergler, S.: Mining WordNet for fuzzy sentiment: Sentiment tag extraction from WordNet glosses. In: Proceedings of the Eleventh Conference of the European Chapter of the Association for Computational Linguistics, pp. 209–216 (2006)

Baroni, M., Vegnaduzzo, S.: Identifying subjective adjectives through Web-based mutual information. In: Proceedings of the Seventh German Conference on Natural Language Processing (2004)

Boucouvalas, A.C.: Real time text-to-emotion engine for expressive Internet communications. In: Being There: Concepts, Effects and Measurement of User Presence in Synthetic Environments, pp. 306–318 (2003)

Castellano, G., Kessous, L., Caridakis, G.: Emotion recognition through multiple modalities: face, body gesture, speech. In: Peter, C., Beale, R. (eds.) Affect and Emotion in Human-Computer Interaction. LNCS, vol. 4868, pp. 92–103. Springer, Heidelberg (2008)

Chayko, M.: Connecting: How we form social bonds and communities in the Internet age. State University of New York Press, Albany (2002)

Derks, D.: Exploring the missing wink: Emoticons in cyberspace. Dissertation. Open University of Netherlands, Leiderdorp (2007)

Di Fiore, F., Quax, P., Vanaken, C., et al.: Conveying emotions through facially animated avatars in networked virtual environments. In: Egges, A., Kamphuis, A., Overmars, M. (eds.) MIG 2008. LNCS, vol. 5277, pp. 222–233. Springer, Heidelberg (2008)

Dimberg, U., Ohman, A.: Behold the wrath: Psychophysiological responses to facial stimuli. Motivation and Emotion 20, 149–182 (1996)

Ekman, P.: Facial expression and emotion. American Psychologist 48, 384–392 (1993)

Esuli, A., Sebastiani, F.: SentiWordNet: a publicly available lexical resource for opinion mining. In: Proceedings of the Fifth International Conference on Language Resources and Evaluation, pp. 417–422 (2006)

Fernback, J., Thompson, B.: Virtual communities: Abort, retry, failure? (2004), `http://www.well.com/user/hlr/texts/VCcivil.html` (Accessed September 8, 2009)

Fridlund, A.J.: The behavioral ecology and sociality of human faces. Review of Personality and Social Psychology 13, 90–121 (1992)

Frijda, N.: The Emotions. Studies in emotion and social interaction. Cambridge University Press, Cambridge (1986)

Frijda, N.: Emotions are functional, most of the time. In: Ekman, P., Davidson, R.J. (eds.) The Nature of Emotion: Fundamental Questions. Oxford University Press, New York (1994)

Harker, L., Keltner, D.: Expressions of positive emotion in women's college yearbook pictures and their relationship to personality and life outcomes across adulthood. Journal of Personality and Social Psychology 80, 112–124 (2001)

Hu, Y., Wood, J.F., Smith, V., et al.: Friendships through IM: Examining the relationship between instant messaging and intimacy. Journal of Computer-Mediated Communication 10(1) (2004)

Izard, C.E.: Human emotions. Plenum Press, New York (1977)

Kamps, J., Marx, M.: Words with attitude. In: Proceedings of Belgian-Dutch Conference on Artificial Intelligence, pp. 449–450 (2002)

Keltner, D., Ekman, P., Gonzaga, G.C., et al.: Facial expression of emotion. In: Davidson, R.J., Scherer, K.R., Goldsmith, H.H. (eds.) Handbook of Affective Science, pp. 415–432. Oxford University Press, Oxford (2003)

Kim, S.-M., Hovy, E.: Automatic detection of opinion bearing words and sentences. In: Proceedings of the International Joint Conference on Natural Language Processing. LNCS. Springer, Heidelberg (2005)

Kim, S.-M., Hovy, E.: Identifying and analyzing judgment opinions. In: Proceedings of the Human Language Technology / North American Association of Computational Linguistics Conference (2006)

Klein, D., Manning, C.D.: Fast exact inference with a factored model for natural language parsing. In: Advances in Neural Information Processing Systems, vol. 15, pp. 3–10. MIT Press, Cambridge (2003)

Leshed, G., Kaye, J.: Understanding how bloggers feel: recognizing affect in blog posts. In: Extended Abstracts of the International Conference on Computer-Human Interaction, pp. 1019–1024 (2006)

Liu, H., Lieberman, H., Selker, T.: A model of textual affect sensing using real-world knowledge. In: Proceedings of the International Conference on Intelligent User Interfaces (IUI 2003), pp. 125–132 (2003)

Lutz, C.: Unnatural emotions. University of Chicago Press, Chicago (1988)

Maglogiannis, I., Vouyioukas, D., Aggelopoulos, C.: Face detection and recognition of natural human emotion using Markov random fields. Personal and Ubiquitous Computing 13(1), 95–101 (2009)

Mihalcea, R., Liu, H.: A corpus-based approach to finding happiness. In: Proceedings of the AAAI Spring Symposium on Computational Approaches to Weblogs (2006)

Miller, G.A.: WordNet: an on-line lexical database. Lexicography (Special Issue) 3(4), 235–312 (1990)

Mishne, G.: Experiments with mood classification in blog posts. In: Proceedings of the First Workshop on Stylistic Analysis of Text for Information Access (2005)

Moilanen, K., Pulman, S.: Sentiment composition. In: Proceedings of the International Conference on Recent Advances in Natural Language Processing, pp. 378–382 (2007)

Montague, R.: Formal philosophy: Selected papers of Richard Montague. Yale University Press, New Haven (1974)

Mulder, M., Nijholt, A., den Uyl, M., Terpstra, P.: A lexical grammatical implementation of affect. In: Proceedings of the Seventh International Conference on Text, Speech and Dialogue, pp. 171–178 (2004)

Nadeau, D., Sabourin, C., De Koninck, J., et al.: Automatic dream sentiment analysis. In: Proceedings of the Workshop on Computational Aesthetics at the Twenty-First National Conference on AI (2006)

Neviarouskaya, A., Prendinger, H., Ishizuka, M.: Analysis of affect expressed through the evolving language of online communication. In: Proceedings of the International Conference on Intelligent User Interfaces (IUI 2007), pp. 278–281. ACM Press, New York (2007a)

Neviarouskaya, A., Prendinger, H., Ishizuka, M.: Textual affect sensing for sociable and expressive online communication. In: Paiva, A.C.R., Prada, R., Picard, R.W. (eds.) ACII 2007. LNCS, vol. 4738, pp. 220–231. Springer, Heidelberg (2007b)

Neviarouskaya, A., Prendinger, H., Ishizuka, M.: User study of AffectIM, an emotionally intelligent Instant Messaging system. In: Proceedings of the Eight International Conference on Intelligent Virtual Agents, pp. 29–36 (2008)

Neviarouskaya, A., Prendinger, H., Ishizuka, M.: EmoHeart: Conveying emotions in Second Life based on affect sensing from text. Advances in Human-Computer Interaction, Special Issue on Emotion-Aware Natural Interaction, Open Access Journal, Hindawi (2010), http://downloads.hindawi.com/journals/ahci/2010/209801.pdf

Olveres, J., Billinghurst, M., Savage, J., Holden, A.: Intelligent, expressive avatars. In: Proceedings of WECC 1998, pp. 47–55 (1998)

Pang, B., Lee, L., Vaithyanathan, S.: Thumbs up? Sentiment classification using machine learning techniques. In: Proceedings of the Conference on Empirical Methods in Natural Language Processing (2002)

Peris, R., Gimeno, M.A., Pinazo, D., et al.: Online chat rooms: Virtual spaces of interaction for socially oriented people. CyberPsychology and Behavior 5(1), 43–51 (2002)

Persson, P.: ExMS: an animated and avatar-based messaging system for expressive peer communication. In: Proceedings of SIGGROUP Conference, pp. 31–39 (2003)

Picard, R.: Affective computing. The MIT Press, Cambridge (1997)

Planalp, S.: Communicating emotion: Social, moral, and cultural processes. Cambridge University Press, Cambridge (1999)

Read, J.: Recognising affect in text using Pointwise-Mutual Information. Master Thesis, University of Sussex (2004)

Read, J.: Using emoticons to reduce dependency in machine learning techniques for sentiment classification. In: Proceedings of the ACL Student Research Workshop at ACL 2005, pp. 43–48 (2005)

Reid, E.M.: Electropolis: Communication and community on Internet Relay Chat. Honours Thesis, University of Melbourne, Australia (1991)

Reilly, J., Seibert, L.: Language and emotion. In: Davidson, R.J., Scherer, K.R., Goldsmith, H.H. (eds.) Handbook of Affective Science, pp. 535–559. Oxford University Press, Oxford (2003)

Rheingold, H.: The virtual community. Revised edition. MIT Press, Cambridge (2002)

Rigas, G., Katsis, C.D., Ganiatsas, G., Fotiadis, D.I.: A user independent, biosignal based, emotion recognition method. In: User Modeling 2007, pp. 314–318. Springer, Heidelberg (2007)

Rivera, K., Cooke, N.J., Bauhs, J.A.: The effects of emotional icons on remote communication. In: Electronic Proceedings of CHI (1996)

Slot, K., Cichosz, J., Bronakowski, L.: Emotion recognition with poincare mapping of voiced-speech segments of utterances. In: Proceedings of the Ninth International Conference on Artificial Intelligence and Soft Computing, pp. 886–895 (2008)

Strapparava, C., Valitutti, A.: WordNet-Affect: an affective extension of WordNet. In: Proceedings of LREC 2004, pp. 1083–1086 (2004)

Strapparava, C., Valitutti, A., Stock, O.: Dances with words. In: Proceedings of IJCAI 2007, pp. 1719–1724 (2007)

Subasic, P., Huettner, A.: Affect analysis of text using fuzzy semantic typing. IEEE Transactions on Fuzzy Systems 9(4), 483–496 (2001)

Tsetserukou, D., Neviarouskaya, A., Prendinger, H., Kawakami, N., Ishizuka, M., Tachi, S.: Affective haptics in emotional communication. In: Proceedings of the International Conference on Affective Computing and Intelligent Interaction, pp. 181–186. IEEE, Amsterdam (2009)

Turney, P.D.: Thumbs up or thumbs down? Semantic orientation applied to unsupervised classification of reviews. In: Proceedings of the Association for Computational Linguistics 40th Anniversary Meeting, pp. 417–424 (2002)
Turney, P.D., Littman, M.L.: Measuring praise and criticism: Inference of semantic orientation from association. ACM Transactions on Information Systems 21(4), 315–346 (2003)
Walther, J., D'Addario, K.: The impacts of emoticons on message interpretation in computer-mediated communication. Social Science Computer Review 19(3), 324–347 (2001)
Wilson, T., Wiebe, J., Hoffmann, P.: Recognizing contextual polarity in phrase-level sentiment analysis. In: Proceedings of HLT/EMNLP Conference, pp. 347–354 (2005)
Wu, C.-H., Yeh, J.-F., Chuang, Z.-J.: Emotion perception and recognition from speech. In: Tao, J., Tan, T. (eds.) Affective Information Processing, pp. 93–110. Springer, London (2009)
Yigit, O.T.: Emoticon usage in task-oriented and socio-emotional contexts in online discussion board. Master Thesis, The Florida State University (2005)
Yu, H., Hatzivassiloglou, V.: Towards answering opinion questions: Separating facts from opinions and identifying the polarity of opinion sentences. In: Proceedings of the Conference on Empirical Methods in Natural Language Processing, pp. 129–136 (2003)

Chapter 10

Interacting with Emotion and Memory Enabled Virtual Characters and Social Robots

Zerrin Kasap[1] and Nadia Magnenat-Thalmann[1,2]

[1] MIRALab, University of Geneva, Switzerland
[2] Nanyang Technological University, Singapore
{zerrin.kasap,thalmann}@miralab.unige.ch
nadiathalmann@ntu.edu.sg

Abstract. Personality and emotion modeling is being used recently in order to create variation in the behaviour of virtual characters and social robots and to make them socially intelligent for more engaged interaction with people. For these characters to be everyday social actors, it is also important that they can remember and maintain their feelings and establish unique emotional interpersonal relationships with people in a long-term manner. Thus they should be equipped with emotion and memory capabilities in order to remember interacted people and past exchanges with them. This chapter gives an overview of theories and applications on emotion, personality and memory enabled intelligent autonomous characters and mentions about our research on creating long-term emotional interactions with a social robot with human-like expression capability.

1 Introduction

Virtual characters and robots that interact with humans just like other humans have forever captured the imagination of science fiction writers and researchers. There has been perfect research on the creation of realistic virtual characters with the improvements in facial and body animation, hair and cloth simulation which has been applied in movies. These characters are more and more used in different application areas from entertainment to training and in such applications the notion of interaction with the user comes to the forefront. It is no longer enough for characters to look like humans but they should also behave and respond like human beings. For the robotics area, there have been many notable efforts recently on creating robotic faces with human like appearance which contribute to the quality of real-artificial interaction with the effect of physical presence. However, creating human-like intelligence for those characters is still a challenging problem.

In this chapter, we give an overview of work on emotion, personality and memory enabled intelligent autonomous characters, both virtual humans and physical robots. Different disciplines such as psychology, cognitive science and artificial intelligence are converged in order to create such human-like behavior. First, in Section 2, we mention about theory and methods that is used to

T. Nishida et al. (Eds.):Modeling Machine Emotions for Realizing Intelligence, SIST 1, pp. 209–224.
springerlink.com

realize personalized and emotional behavior for virtual characters and robots. Later, in Section 3, we present a literature review of episodic memory from the neurological, psychological and AI perspective and then give examples of episodic memory enabled virtual characters and social robots. Finally, we mention about our current research and future research plans for creating long-term emotional interactions with a social robot with human-like expression capability.

2 Personality, Emotions and Social Relationships

2.1 Emotions

Emotions have proven effects on cognitive processes such as action selection, learning, memory, motivation and planning. Our emotions take role as motivational factors for our decisions and effect our actions. We can see emotions as one of the important mechanisms to control the behaviour of virtual characters, both for the reason of giving them a personality as well as automatically producing animations by simulating the internal dynamics of the virtual characters.

Personality and emotions have been widely researched in the field of psychology and comprehensive computational models were developed which were later utilized by AI researchers for the modeling of emotions and personality in life-like characters. Gratch and Marsella [1] define two methods for emotion modeling: communicative-driven methods and simulation-based methods. Communicative-driven methods treat emotional displays as a means of communication. These systems does not have an internal calculation of emotion but selects a display of emotion based on the current state of interaction and usually encoded via some scripting language. Simulation-based approaches attempt to model the impact of events on internal emotion dynamics focusing on the cognitive function of emotion. Emotional displays are tied to the virtual character's emotional state rather than being triggered by their communicative function.

The first group of systems usually use basic emotions introduced by Ekman [2] with six universally accepted emotion labels: fear, disgust, anger, sadness, surprise and joy. Another categoric emotion model is the Plutchik's emotion spectrum [3]. Plutchik defines eight emotions in an emotion spectrum similar to a color spectrum and new emotion labels are produced by the way of mixing different emotions (e.g. disappointment = sadness + surprise).

Second group of systems models the appraisal of events in the environment and their effects on the internal emotional state. Examples of these models are of Lazarus [4], Roseman [5], Scherer [6] and most popularly used OCC model of Ortony, Clore and Collins [7]. In OCC model, agent's concerns in an environment are divided into goals (desired states of the world), standards (ideas about how people should act) and preferences (likes and dislikes) and twenty-two emotion labels are defined. Later, Ortony himself found the twenty-two distinct emotion types too complex for the simulation of believable characters and decreased these labels to twelve, six being positive (joy, hope, relief, pride, gratitude, love) and six being negative (distress, fear, disappointment, remorse, anger) [8]. In OCC, five phases are defined for the appraisal of events, agents and objects in the environment [9] : (1) Classification, where the character evaluates an event, action or object, resulting in information on what emotional categories are affected, (2) Quantification; where emotional intensities of the affected emotional categories are decided (3) Interaction, which means how the emotional state will interact with the current emotional categories of the character (4) Mapping is how OCC emotions are represented by a lower number of different expressions and (5) Expression means how these will be expressed as facial expressions or behaviours. Classification is done based on the knowledge available, like what are the goals of the character, what are the standards for other people's behaviours and attitudes towards objects.

2.2 Mood

Moods are average internal affective states of an individual. Gratch and Marsella [10] differentiate between mood and emotions based on three criteria: temporal, expression and cause. Moods last longer than emotions and they are not associated with a specific event. In other words, emotions modulate actions while moods modulate cognition. The relation between mood and emotions is two-way. Mood affects the appraisal of events and decides which emotion will be triggered and with what intensity. For example, if a person is in an anxious mood, he/she will get more easily disappointed by bad events and with higher intensity. Emotions can also cause a particular mood to occur. For example, a person in a bored mood can become in a more positive mood after some emotional appraisal from the environment. Moods are usually represented with continuous dimensions rather than being discrete labels. Dimensional models such as activation-evaluation [11] and Mehrabian's PAD model [12] is commonly used for the modeling of moods. Sometimes it is also represented as one-dimensional like being in a good or bad mood.

2.3 Personality

Personality influences the way people perceives their environment, affects their behaviors and actions and distinguish one person from the others. In [10], two diferences are mentioned between emotions and personality: duration and focus. Personality is constant and like mood it is not specific to particular events. For example, a person who is stable will have the tendency to behave in a less emotional way in difficult situations. Although there is no universally accepted theory of personality, Five Factor Model or OCEAN model [13] is most widely used in the personality simulation of computer characters. According to this model, personality of a person can be defined according to five different traits [14]:

- **Openness (O):** open to experience new things, being imaginative, intelligent and creative
- **Conscientiousness (C):** indicates responsibility, reliability and tidiness. Conscientious people think about all the outputs of their behaviors before taking action and take responsibility.
- **Extravertion (E):** outgoing, sociable, assertive and energetic to achieve his/her goals.
- **Agreeableness (A):** trustable, kind and cooperative considering other people's goals and is ready to surrender his own goals.
- **Neuroticism (N):** anxious, nervous, prone to depression and lack of emotional stability.

Usually, a character is represented as a combination of these traits possibly with emphasize on one of them. Although this static trait-based model of personality does not reflect the complexity of real human behavior, it has been widely accepted to be used in computational models because of its simplicity.

2.4 Social Relationships

Another very important factor that shapes our every day emotional reactions is the status of our interpersonal relations with other people. This concept becomes important when the conversational partners are coming together multiple times forming a long-term interaction. In [15], Bickmore mentions about various models of relationship based on social psychology, such as dyadic, provision, economic, dimensional and stage models.

- **Dyadic models** define relationship as the interdependency between two people such that a change in the state of one will produce a change in the state of the other.
- **Provision models** are based on what one person provides for the other.
- **Economic models** such as social-exchange theory, model relationships in terms of costs and benefits
- **Stage models** assume that relationships go through a fixed set of stages.
- **Dimensional models** attempt to abstract a given relationships characteristics to a point in a small dimensional Euclidean space.

From these models, dimensional models are more widely used for the modeling of life-like characters. The most commonly used dimensions are power and social distance [16], where power refers to the ability of one individual to control the resources of another and social distance refers to how close two person and what is the level of friendship between them. Argyle [17] makes a similar definition and proposes a model based on two dimensions:

friendliness and dominance. In [18], Bickmore and Cassell mentions about two dimensions, self-disclosure and trust, which are important in health and sales applications. In [19], three dimensions are considered: familiarity (degree to which people have exchanged information), solidarity (how similar two people are to eachother in their interests, beliefs) and affect (the degree of liking).

2.5 Virtual Characters and Social Robots with Emotional Interaction Capability

Personality, mood and emotions are used in many applications as a three layer model. For example, Krshirsagar et. al. [20] uses OCC model for emotions and OCEAN model for personality. Mood is represented as good, bad or neutral and Bayesian Belief Networks are used in order to model the uncertainty in mood change based on different personality. In [21], a generic model is described for emotional communication, where different theory of emotion, mood and personality can be applied. ALMA (A Layered Model of Affect) [22] uses a mood model with three dimensions based on the PAD model of Mehrabian [12] where OCEAN personality decides the initial mood of the virtual characters. In the embodied conversational agent MAX's emotion system [23], emotion and mood is represented as an orthogonal system where the mutual interaction between them is explained with a mass-spring system. Kismet [24] is one of the first examples of social robots with an emotion system. In Kismet's emotion system, emotional appraisals from the environment are represented with arousal, valence and stance dimensions and these are converted to distinct affective processes (e.g. anger, fear, etc). iCat developed by Philips is a cartoon-like robot cat with facial expressions that can play chess with a human opponent. Each move of the user is interpreted as an emotional input and effects iCat's emotional system and its response [25]. These applications does not take into account the effect of interpersonal relationships on the emotion and mood level. One of the first applications that attempt to model interpersonal relationships is [18]. The authors introduce the term "relational agents" and mention about a relationship model based on trust and self-disclosure. Predinger [26] mentions about how relationships evolve over time as a consequence of affective interaction histories with other agents based on OCC appraisal model. In BASIC [27], a socially intelligent agent is developed considering factors such as emotion, mood, personality, memory of relationships with other agents and social cognitive factors. Gockley [28] describes an affective system for a socially-interactive virtual robot based on emotion, mood and attitudes towards users where attitudes are represented with dimensions like familiarity and affect.

3 Episodic Memory

In order to believe that an embodied character is intelligent, it should remember past knowledge and refer to that knowledge where necessary. The character should have a memory that can store and retrieve past experiences so that it can reason about them and behave appropriately. Besides the reasoning factor, the existence of a real user in the interaction loop makes the use of memory necessary in order to increase the believability of the character from the perspective of the user. The moment the character does not remember a past exchange with the interactor, he/she will feel no more engaged in the interaction losing his/her trust towards the character.

3.1 Background

Episodic memory was first described by Endel Tulving [29] who is a neuroscientist. According to his work, memory is composed of long-term and short-term (working) memory and long-term memory is divided into declarative and non-declarative memories (Fig 1). Non-declarative memory is also called procedural memory which is related with learning skills like riding a bicycle or playing a guitar which requires the storage and retrieval of sensory-motor skills. Declarative memory is composed of episodic and semantic memory. Even though both episodic and semantic memory is related with the acquisition and retrieval of knowledge, they differ in many aspects. Episodic memory represents our experiences as points on a timeline associating each event with a point in time. In [45], episodic memory is defined as a "neurocognitive (brain/mind) system, uniquely different from other memory systems that enable human beings to remember past experiences". Semantic memory is derived from episodic memory and it is a structured representation of learned facts and concepts. For example, remembering what you have done last summer is an episodic memory and your knowledge about a city is semantic memory. Episodic memory stores events in a temporally manner and it is based on the experiencer's view.

Tulving [29] defines the following properties of episodic memories:

> **Autobiographical** - a person remembers episodes from his/her own perspective.
> **Autonoetic** - remembering entails the conscious re-experience of past memories, but the retrieved memories are distinguished from the perception of the persons current state.
> **Temporally annotated** - the person has a sense of the time when an episode has occurred.
> **Imperfect** - the memory is incomplete and can have errors.
> **Activated** - exposure frequency and recency affect the speed and probability of recall.
> **Primed** - recall occurs more quickly when it is primed by repetition or by recall of related information or similar state.
> **Forgetting** - memory performance declines with time or intervening events (....) (Tecuci, [34], p. 9)

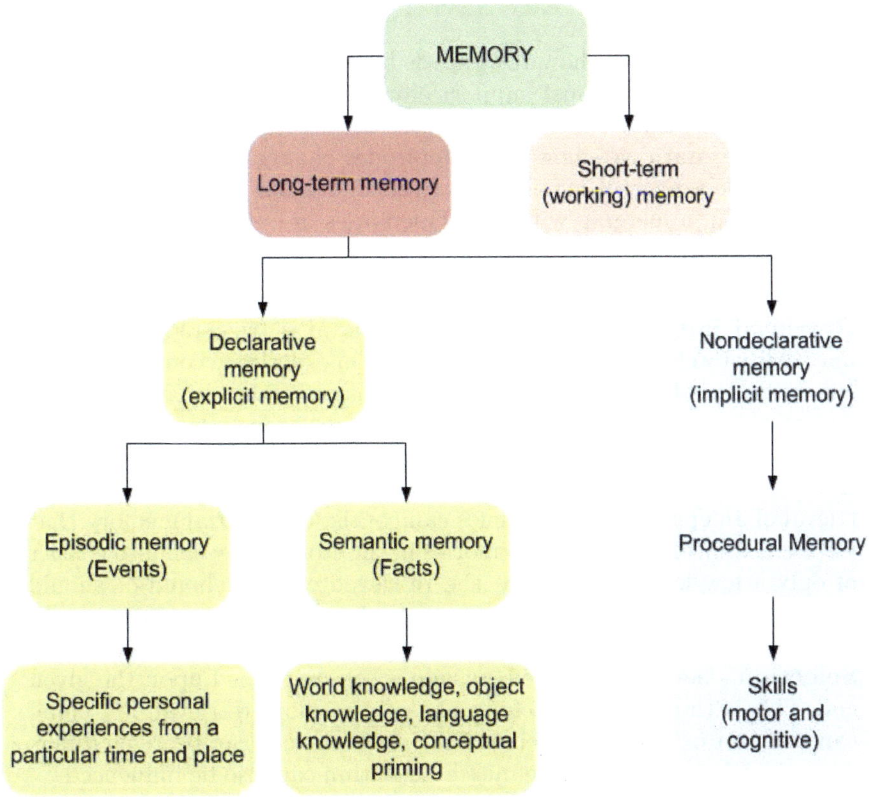

Fig. 1. Classification of Memory Types

3.2 How Does Episodic Memory Work?

Episodic memory process is composed of three phases: encoding, storage and retrieval [29]. Nuxoll [30] (p. 7-8) explains these stages as below:

> **Encoding**
> **Encoding initiation** - when an episode is recorded. (....) Alternatively, a new episode might be recorded whenever the agent takes an action in the world, whenever there is a significant change in its sensing, or when something unexpected occurs.
> **Episode determination** - what information is stored in an episode. (....) It is at this stage that the episodic memory system must select what is "important" about the agent's current state that should be stored for possible later retrieval.
> **Feature selection** - which features in an episode will be available for matching against a cue during the retrieval phase. The selection of features may include the entire episode or as little as one or two key features. (....)

Storage
Episode structure - how the encoded episode is stored. (....) In particular, the structure must support efficient and stable storage as well as expedient episode retrieval.
Episode dynamics - how stored episodes change over time. (....). It may include a form of memory decay such as a loss of detail, a decline in activation, a merging with other memories or outright removal from the episodic store.

Retrieval
Retrieval initiation - how memory retrieval is triggered. The agent may deliberately initiate memory retrieval in order to accomplish a specific subtask where it deems that previous experience may be useful. Episodic retrieval may also be triggered spontaneously by the presence of strongly familiar elements or unexpected sensing.
Cue determination - how data are selected or created to cue the retrieval of an episode. Memory cues can consist of a partial memory that is matched directly to other memories in the store. They can also consist of only a few key features that the retrieved episode should or should not contain. Finally, a memory cue can also contain meta-information that is relative to the current state or the most recent retrieval (....).
Selection - how the retrieved episode is selected based upon the given cue. This is the critical matching phase of memory retrieval. Matching can be exact or partial. Different features of the cue can be given different weight in the match. The match algorithm can also be influenced by the agent's current sensing and internal state. Multiple memories can be retrieved or just a single episode.
Retrieval - what aspects of the episode(s) are retrieved and how they are represented. (....)
Retrieval meta-data - what meta-information about the retrieved episode is available. Meta-data that could be available includes: information about the strength of the match between the episode and the cue; temporal information describing when the episode occurred; and information about when or how often the episode has been retrieved in the past.

Even though episodic memory has been searched well from the neurological perspective, there is a lack of extensive computational models of episodic memory from the psychology literature. In the next section, we will mention about the existing computational models of episodic memory and give some examples of virtual characters and robots with episodic memory capability.

3.3 Virtual Characters and Social Robots with Episodic Memory

One of the first computational models of memory was introduced by Schank [31]. He proposed various memory structures such as "scripts (sequences of actions), scenes (group of actions that happen at the same time) and memory organization packets (MOPs) (ordered set of scenes directed toward the achievement of a goal not inferable from individual scenes)" as defined by [34] (p.13). His model of dynamic memory was the basis for the Case-Based Reasoning (CBR) from the AI literature which is a technique to solve problems based on the solutions of similar past problems [32]. Case-based reasoning is very close to the idea of episodic memory, however they also differ in significant ways [33]. For example, the reasoning mechanism in CBR is included in the method itself and realized with case reuse and adaptation. However, episodic memory can be integrated with other applications and reasoning mechanisms. It is more general and application independent which means that once it is constructed, it could be used for multiple purposes such as action selection, learning, problem solving and storytelling.

SOAR and ACT-R are the two general purpose cognitive architectures that have declarative and procedural memory but episodic memory was relatively ignored. SOAR architecture has recently been extended with a robust, general purpose episodic memory functionality [30]. Episodic memory is added to this architecture by the way of storing everything in working memory except the working memory elements with very low activation. Quering is done by creating a structure in the cue area of working memory. Episode that is the best match to the cue is found and deposited in the retrieved area of working memory. Tecuci et. al. [34] presents a generic episodic memory module where an episode is represented in terms of context, content and outcome. While SOAR-EM uses feature match algorithm for computing the similarity between the cue and a prior episode, they use a semantic matcher.

Since the attention of the AI community to the episodic memory has arised very recently, AI approaches that aims to integrate computational models of episodic memory into the virtual characters and social robots are very limited. Thus creating life-long characters that has the capability of long-term interaction in virtual and real environments is a challenging and state-of-the-art topic.

One of the first example of agents that can remember past experiences is Vere and Bickmore's "Basic Agent" [35] where all perceptions are recorded as sense events and sentences are recorded as inform events. Pedagogical training agent Steve [36] uses an episodic memory structure in order to rationalize his own actions during an after-action review with the trainees. Ho et. al. [37] mentions about autobiographic memory and integrates it with IVAs (Intelligent Virtual Agents) with story-telling capabilities where agents can remember significant past experiences and reconstruct their life stories from these experiences, in an emotion-driven planning architecture. It has been used in the FearNot! application which was developed to tackle and help to reduce bullying problems in schools. In this application, a child sees one episode where a bullying situation occurs and after each episode the child interacts with the victim and discusses the problems that arise and propose coping stragtegies. The suggested strategy is taken into account by the victim and applied in the next scenario. In [38], Brom et. al. proposes a virtual role-playing game (RPG) actor with a full episodic memory. The authors mention that in RPGs, it is important that non-player characters (NPCs) can tell believable stories about themselves during interaction with the player characters (PCs). The story shouldn't be predefined but should be constructed dynamically based on what NPCs are experiencing. Therefore, there is a need for a full episodic memory and a linguistic module transferring the outcome of the memory to natural language.

In [39], an episodic memory system is presented for the humanoid robot ISAC (Intelligent Soft Arm Control). ISAC's cognitive architecture is designed to be able to learn appropriate behaviour through interaction with its environment. ISAC is also equipped with an emotion system and episodes in the memory is associated with an emotional value. In an experiment, a person enters the room, and yells "Fire!". ISAC retrieves the episode associating the current situation with a highly emotionally salient past experience, and stop its current task and warns the experimenters to leave the room based on its past experience. Jockel et. al. [40] investigate the use of episodic memory (EPIROME) in the multi-modal service robot TASER which serves drinks to guests in an office environment. In EPIROME, episodic memories are considered as sequences of events and each event carries a time information and can be assigned to one of the three major event classes: perceptual, command and executive. Mirza et. al. [41] uses the term interaction histories instead of episodic memory, which is defined as "the temporally extended, dynamically constructed, individial sensorimotor history of an agent situated and acting in its environment, including the social environment, that shapes current and future action". In their study, humanoid robot Kaspar2 learns to engage with a human partner in a simple "peekabo" interaction game based on past interactions. In [42], anticipatory behaviour of a navigation robot is computed based on the episodic-memory. In this study, robot learns to localize itself using past sensory and behavioral information in a partially observable environment.

4 Case Study: Long-Term Emotional Interaction with a Social Robot

In this section we mention about our research on creating emotion and memory-enabled socially interactive robots. Our goal is to create long-term emotional interaction between humans and social robots and establish social relationships based on past knowledge. Episodic memory is the core structure for that. We believe that the purpose of having an episodic memory for artificial characters is twofold: (1) Having a sense of self (2) Having a sense of others. Virtual characters should remember their relationships with the other virtual characters, remember important and related events in their memory, reason on them and talk about past experiences from their own perspective. Construction of episodic memory is also very much related with personality and emotion of the characters since emotionally salient events are remembered more and have more effect on the characters' life. The challenge is to

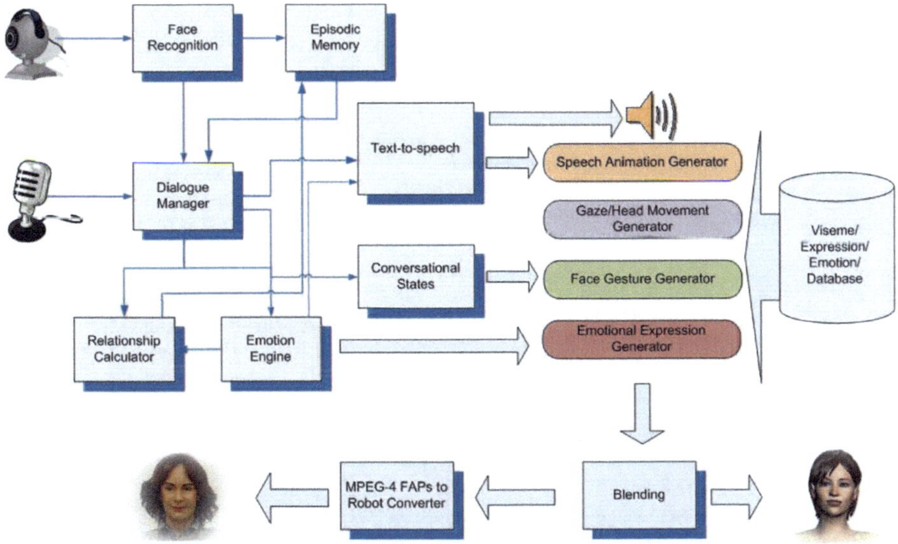

Fig. 2. Interaction Architecture

develop an appropriate memory model that considers what information will be stored in the memory, how it is retrieved and how long it will stay in the memory, based on the importance and relevance of the events, as well as based on the personality and emotion of the virtual character.

Fig 2 shows our system's interaction architecture. In this architecture emotions, mood and personality of the robotic character are derived from an emotion engine that we have developed in [43]. An artificial character with this emotion engine has a personality much like a human being has, and is also able to emotionally respond to the user during a conversation. Emotion engine is linked with an episodic memory model that uses the memory of past interactions to build long-term relationships between the robot and human users. As a relationship model, we use Argyle's model [17] based on two dimensions; friendliness and dominance. The relationship with a user affects the emotional reactions of the virtual character. For example if the user is saying something bad happened to him/her and the character has positive impressions of the person, the resulting emotion will be sorry-for but if the person is not liked, the character will experience the gloating emotion. Relationship update is based on the cumulative evaluation of the individual's overall interaction with the user. For each interaction, positive and negative impulses from user help to construct a relationship between the interactants. In order to associate events with different users, a face detection and recognition module is integrated into our system. According to the scenario two students interacts with our virtual robot Eva, one playing a good student and the other playing the role of a difficult student. The face recognition module automatically detects a new person whenever somebody arrives and Eva

	Good Student	Difficult student
First Interactions	*Eva:* Nice to meet you Maher. I would like to start with a few simple concepts first. Do you know about different landforms on the Earth? *Student:* No, not really. I am eager to learn. *Eva:* Ok, today we will talk about three types of landforms: cape, strait, and gulf. Then we will have a small test to check if you understood the concepts. Is it ok for you? *Student:* Yes, fine for me. *Eva:* Good. Let's start with the cape. A cape is a pointed piece of land that sticks into the sea, ocean, lake, or river. A good example is the Cape of Good Hope to the south of Cape Town. Do you know what a strait is? *Student:* No.	*Eva:* Nice to meet you Zerrin. I will start with a few simple concepts first. Do you know about different landforms on the Earth? *Student:* I don't know anything about them. Boring topics! *Eva:* Ok, let me explain some of them to you. It is not that boring, trust me. *Student:* If you say so. *Eva:* Today we will talk about three types of landforms: cape, strait, and gulf. Then we will have a small test to check if you understood the concepts. Is it ok for you? *Student:* You teachers always want to take tests. *Eva:* It is important for you to learn these concepts. Let's start with the cape. A cape is a pointed piece of land that sticks into the sea, ocean, lake, or river. A good example is the Cape of Good Hope to the south of Cape Town. Do you know what a strait is? *Student:* No.
Second Interactions	*Eva:* Hi Maher, great to see you! *Student:* Hi Eva! *Eva:* You were very good in our last class. *Student:* Thank you. You are a nice teacher too. *Eva:* You really think so? Thank you. Are you ready to learn more things today? *Student:* I don't feel well today. Maybe it is better if we continue later on. *Eva:* Oh, I am really sorry to hear that. Ok. See you another time. Bye. *Student:* Bye.	*Eva:* Hi Zerrin! You are back again? I thought you were not very interested in geography. *Student:* Hi Eva! I am sorry for last time. I will try my best from now on. *Eva:* Well, I hope so. Ok, Let's start the lesson then. *Student:* I don't feel well today. Can we continue later on? *Eva:* I hope this is the truth. See you later then. Bye. *Eva:* Bye.

Fig. 3. Example dialogues of Eva showing two interaction sessions with good and difficult students ([42] p.27)

greets the student. Recognition of a new user starts a new interaction session which is a new episode in the episodic memory. With the good student Eva stays calm and polite while with the difficult student she is more aloof in her responses. After some time, the two students have another interaction with Eva. Face recognition module recognizes the students automatically and Eva remembers their faces. She also remembers past relationship levels with the students and updates her emotional state according to the relationships and answers accordingly. Fig 3 shows snippets from the four dialogues with good and difficult students in the first and second interactions. For more details of the components and the scenario, we refer to [43].

As a final output, emotion engine and other behaviour generators[1] generates facial animation for virtual characters. In our research, we use a robotic head developed by Hanson Robotics [44] that has extremely realistic looking skin giving it almost a lifelike appearance to capture the mechanics of the face. We map the facial animation parameters of a virtual face to the robotic face and show how the facial animation of a virtual character can be used to generate expressions on the robotic face. The facial animation is described via MPEG-4 Facial Animation Parameters (FAPs). The FAPs are mapped to the movements of the servo-motors that control the facial expressions of the robotic face, thus transferring the facial animation from virtual to the robot. In Fig 4, we show a screen-shot from our FAPtoRobotConverter application and expression applied on the robotic head.

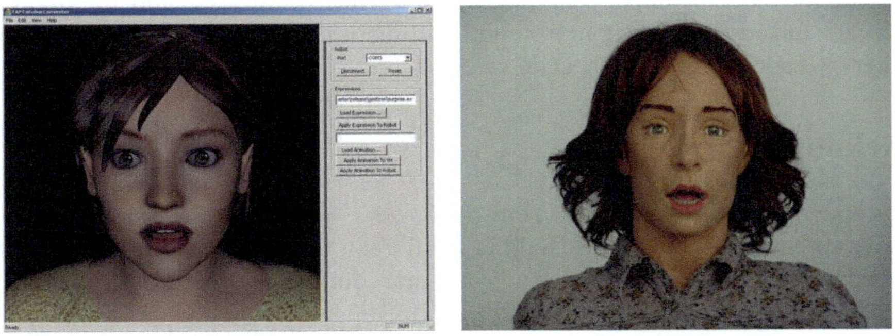

(a) FAPtoRobotConverter Application (b) Robotic head

Fig. 4. Emotional expression applied on the virtual character and robotic head

Future Research Plans: Our future goal is twofold. First, we would like to improve our episodic memory model so that the character can not only remember overall relationships after each interaction session but also remember specific events in its memory, reason on them and talk about past experiences from his/her own perspective. In the episodic memory process it is important to decide on the following questions:

- When an episode will be recorded?
- What information will be stored?
- Which cues can be chosen to store the episode for use in the retrieval process?
- How long an episode will stay in memory, how does it change over time?
- What triggers the retrieval of the episode?
- How does the current situation linked to the episode for retrieval?

[1] Speech and face gesture generators are out of the scope of this book chapter.

Episodic memory and emotion engine should be linked with the dialogue manager so that the virtual character can give appropriate responses based on the information received from memory and current emotional state. Besides remembering the past events related with the current dialogue context and giving appropriate responses, we expect the character to tell his/her own experiences by combining multiple entries in the episodic memory so that he/she can answer questions like "What did you today, did you have a nice day?" and start telling you how his/her they has passed, which people he/she has interacted with and which of them he/she liked etc.

Our second goal for our system is to improve the behavior generators so that they can be much more natural and un-repetitive. For more expressive and natural behavior generation, we plan to use PAD (pleasure-arousal-dominance) driven facial expression and head/gaze behavior generators since pleasure-arousal-dominance factors can be well-linked with the quality of the movements such as positive-negative, energetic-unenergetic and dominant-submissive. For example, emotional levels such as arousal and dominance can be linked with gaze movement parameters such as saccade duration, inter-saccade interval or saccade velocity.

Acknowledgement

We like to thank Maher Ben Moussa for his contribution to the robotic facial expression software and Nedjma-Cadi Yazli for creating facial expression database for the robot. This study is partly funded by the EU Project INDIGO (IST-FP6 045388).

References

[1] Gratch, J., Marsella, S.: Lessons from emotion psychology for the design of lifelike characters. Applied Artificial Intelligence 19(3-4), 215–233 (2005)
[2] Ekman, P.: Emotion in the Human Face. Cambridge University Press, New York (1982)
[3] Plutchik, R.: A general psychoevolutionary theory of emotion. In: Plutchik, R., Kellerman, H. (eds.) Emotion: Theory, research, and experience, vol. 1, pp. 2–33 (1980)
[4] Lazharus, R.: Emotion and Adaptation. Oxford University Press, Oxford (1991)
[5] Roseman, I., Spindel, M., Jose, P.: Appraisals of emotion-eliciting events: Testing a theory of discrete emotions. Personality and Social Psychology 59(5), 899–913 (1990)
[6] Scherer, K.R., Shorr, A., Johnstone, T.: Appraisal processes in emotion: theory, methods, research. Oxford University Press, Canary (2001)
[7] Ortony, A., Clore, G.L., Collins, A.: The Cognitive Structure of Emotions. Cambridge University Press, Cambridge (1988)
[8] Ortony, A.: On making believable emotional agents believable. In: Trappl, R., et al. (eds.) Emotions in Humans and Artifacts, pp. 189–212 (2003)

[9] Bartneck, C.: Integrating the occ model of emotions in embodied characters. In: Proceedings of the Workshop on Workshop on Virtual Conversational Characters: Applications, Methods, and Research Challenges, Melbourne (2002)
[10] Vinayagamoorthy, V., Gillies, M., Steed, A., Tanguy, E., Pan, X., Loscos, C., Slater, M.: Building expression into virtual characters. In: Eurographics Conference State of the Art Reports (2006)
[11] Whissel, C.M.: The dictionary of affect in language (1980)
[12] Mehrabian, A.: Pleasure-arousal-dominance: A general framework for describing and measuring individual differences in temperament. Current Psychology 14, 261–292 (1996)
[13] Mccrae, R.R., John, P.: An introduction to the five-factor model and its applications. Journal of Personality 60, 175–215 (1992)
[14] Hampson, S.: State of the art: Personality. The Psychologist 12(6), 284–290 (1999)
[15] Bickmore, T.W., Picard, R.W.: Establishing and maintaining long-term human-computer relationships. ACM Trans. Comput.-Hum. Interact. 12(2), 293–327 (2005)
[16] Svennevig, J.: Getting acquainted in conversation. John Benjamins, Philadephia (1999)
[17] Argyle, M.: Bodily communication, 2nd edn. Methuen and Co Ltd (1998)
[18] Bickmore, T.W., Cassell, J.: Relational agents: A model and implementation of building user trust. In: Proceedings of SIGCHI (2001), pp. 396–403 (2001)
[19] Bickmore, T.W., Cassell, J.: Social dialogue with embodied conversational agents. In: van Kuppevelt, J., Dybkjaer, L., Bernsen, N. (eds.) Advances in Natural, Multimodal Dialogue Systems, pp. 256–284. Kluwer Academic, Dordrecht (2005)
[20] Kshirsagar, S., Magnenat-Thalmann, N.: A multilayer personality model. In: ACM International Conference Proceeding Series, vol. 24, pp. 107–115 (2002)
[21] Egges, A., Kshirsagar, S., Magnenat-Thalmann, N.: Generic personality and emotion simulation for conversational agents. Computer Animation and Virtual Worlds 15(1), 1–13 (2004)
[22] Gebhard, P.: Alma - a layered model of affect. In: Proceedings of the Fourth International Joint Conference on Autonomous Agents and Multi Agent Systems, pp. 29–36 (2005)
[23] Becker, C., Kopp, S., Wachsmuth, I.: Simulating the emotion dynamics of a multimodal conversational agent. In: André, E., Dybkjær, L., Minker, W., Heisterkamp, P. (eds.) ADS 2004. LNCS (LNAI), vol. 3068, pp. 154–165. Springer, Heidelberg (2004)
[24] Breazeal, C.: Emotion and sociable humanoid robots. International Journal of Human-Computer Studies 59, 119–155 (2002)
[25] Leite, I., Pereira, A.: Icat, the affective chess player. In: Petta, P., Martinho, C. (eds.) Proceedings of the systems demonstrations. Second international conference on Affective Computing and Intelligent Interaction (ACII 2007), pp. 29–33 (2007)
[26] Predinger, H., Ishizuka, M.: Evolving social relationships with animate characters. In: Proceedings of the Symposium of the AISM Convention and Animationg Expressive Characters for Social Interactions, pp. 73–78 (2002)
[27] Romano, D., Sheppard, G., Hall, J., Miller, A., Ma, Z.: Basic: A believable, adaptable socially intelligent character for social presence. In: PRESENCE 2005, The 8th Annual International Workshop on Presence, London, UK (2005)

[28] Gockley, R., Simmons, R., Forlizzi, J.: Modeling affect in socially interactive robots. In: The 15th IEEE International Symposium on Robot and Human Interactive Communication, ROMAN 2006 (2006)
[29] Tulving, E.: Elements of Episodic Memory. Claredon Press, Oxford (1983)
[30] Nuxoll, A.: Enhancing Intelligent Agents with Episodic Memory. PhD thesis, The University of Michigan (2007)
[31] Schank, R.: Dynamic Memory. A Theory of Reminding and Learning in Computers and People. Cambridge University Press, Cambridge (1982)
[32] Kolodner, J.: Case-based reasoning. Morgan Kaufmann, San Mateo (1993)
[33] Laird, J.: Episodic memory vs. case-based reasoning. In: Invited talk at the Sixth International Conference on Case-Based Reasoning, ICCBR (2005)
[34] Tecuci, D.: A Generic Memory Module for Events, Phd Dissertation. Computer Sciences Department, The University of Texas at Austin (August 2007)
[35] Vere, S., Bickmore, T.: A basic agent. Computational Intelligence 4, 41–60 (1990)
[36] Rickel, J., Johnson, W.: Animated agents for procedural training in virtual reality: Perception, cognition, and motor control. Applied Artificial Intelligence 13(4-5), 343–382 (1999)
[37] Ho, W.C., Dautenhahn, K.: Towards a narrative mind: Creating coherent life stories for believable agents. In: Prendinger, H., Lester, J.C., Ishizuka, M. (eds.) IVA 2008. LNCS (LNAI), vol. 5208, pp. 59–72. Springer, Heidelberg (2008)
[38] Brom, C., Pesková, K., Lukavský, J.: What does your actor remember? towards characters with a full episodic memory. In: International Conference on Virtual Storytelling, pp. 89–101 (2007)
[39] Dodd, W., Gutierrez, R.: The role of episodic memory and emotion in a cognitive robot. In: IEEE International Workshop on Robot and Human Interactive Communication, Roman, pp. 692–697 (2005)
[40] Jockel, S., Westhoff, D., Zhang, J.: Epirome - a novel framework to investigate high-level episodic robot memory. In: IEEE International Conference on Robotics and Biomimetics, pp. 1075–1080 (2007)
[41] Mirza, N., Nehaniv, C., Dautenhahn, K., te Boekhorst, R.: Developing social action capabilities in a humanoid robot using an interaction history architecture. In: 8th IEEE-RAS International Conference on Humanoid Robots, pp. 609–616 (2008)
[42] Endo, Y.: Anticipatory robot control for a partially observable environment using episodic memories. In: ICRA 2008, pp. 2852–2859 (2008)
[43] Kasap, Z., Ben Moussa, M., Chaudhuri, P., Magnenat-Thalmann, N.: Making them rememberemotional virtual characters with memory. IEEE Computer Graphics and Applications 29(2), 20–29 (2009)
[44] Hanson, D., Olney, A., Zielke, M., Pereira, A.: Upending the uncanny valley. In: Proceedings of National Conference on Artificial Intelligence, AAAI 2005 (2005)
[45] Tulving E.: Episodic memory: from mind to brain. Annual Rev. Psychol. 53, 1–25 (2002)

Author Index

Editors

Toyoaki Nishida is Professor at Department of Intelligence Science and Technology, Graduate School of Informatics, Kyoto University. He received the B.E., the M.E., and the Doctor of Engineering degrees from Kyoto University in 1977, 1979, and 1984, respectively. His research centers on Artificial Intelligence and Human Computer Interaction. Toyoaki Nishida proposed Social Intelligence Design as a new research field and has been working on it, as well as co-chairing a series of international workshops on Social Intelligence Design. Social Intelligence Design as a research field attempts to integrate understanding and designing social intelligence. Engineering aspects of Social Intelligence Design involve design and implementation of systems and environments, ranging from group/team oriented collaboration support systems that facilitate common ground building, goal-oriented interaction among participants, to community support systems for supporting large-scale online-discussion. Scientific aspects involve cognitive and social psychological understanding of social intelligence, attempting to provide a means for predicting and evaluating the effect of a given communication medium on the nature of discussions, interaction dynamics, and conclusions. Nishida places a particular emphasis on the use of conversational communications in the knowledge process and introduces a computational framework of conversational knowledge process. In 2007, he proposed Conversational Informatics as a field of research aiming at investigating human conversational behaviors as well as designing conversational artifacts that can interact with people in a conversational fashion. Conversational Informatics attempts to establish a new technology consisting of environmental media, embodied conversational agents, and management of conversational contents, based on the foundation of Artificial Intelligence, Pattern Recognition, and Cognitive Science. The major application areas of Conversational Informatics involve knowledge management and e-learning. Conversational Informatics covers a broad field of research encompassing linguistics, psychology and human-computer interaction and interdisciplinary approaches are truly important. Nishida emphasized the engineering aspects such as conversational content acquisition, conversation environment design and quantitative conversational modeling. Recently, Nishida attempts to base Social Intelligence Design and Conversational Informatics on deeper levels of human cognition and embodiment in search for Primordial Knowledge Models.

Professor L.C. Jain is a Director/Founder of the Knowledge-Based Intelligent Engineering Systems (KES) Centre, located in the University of South Australia.

His interests focus on the applications of novel techniques such as knowledge-based systems, virtual intelligent systems, defence systems, intelligence-based medical systems, e-Education and intelligent agents.

Dr. Colette Faucher is an Associate Professor at Polytech'Marseille, a School of Industrial Engineering located in Marseilles, France. She has been teaching since 1990. She received her PhD in 1991 and her Habilitation to Direct Research in 2002. Her research interests include Knowledge Representation and Reasoning. She is particularly interested in Classification Reasoning and Concept Formation from a psychological point of view. She has been the Chair of many conferences and a member of program committees of various conferences. She is a member of several editorial boards. She is very interested in Cognitive Psychology and all her research is based on psychological works. She is presently working on Causal Reasoning and the modelling of Emotions in Knowledge-Based Systems.

GPSR Compliance

The European Union's (EU) General Product Safety Regulation (GPSR) is a set of rules that requires consumer products to be safe and our obligations to ensure this.

If you have any concerns about our products, you can contact us on ProductSafety@springernature.com

In case Publisher is established outside the EU, the EU authorized representative is:

Springer Nature Customer Service Center GmbH
Europaplatz 3
69115 Heidelberg, Germany

Batch number: 10614744

Printed by Printforce, the Netherlands